AN ILLUSTRATED GUIDE TO CAT DISSECTION

By

Clifford E. Keeney, Ph.D.

Dissection Photography by Clifford S. Keeney

MP

Morton Publishing Company
295 West Hampden, Suite 104
Englewood, Colorado 80110

Illustrated by Susan Strawn and Richard Fritzler

ISBN: 0-89582-080-3

Contents

Preface

Most students enrolled in undergraduate courses in human or mammalian anatomy and physiology are studying the health-related sciences, such as nursing, physical education, or the newly emerging programs in health and corporate fitness. Their primary interest is in the structure and function of the human organism, and indeed, this is a valid interest, since they will be nursing, teaching, and prescribing exercise programs for humans, not cats. Human cadavers, however, are extremely expensive and difficult to obtain. In my state, for example, an institution must have a legally established morgue in order to obtain cadavers. Such reasons have made the common housecat the mammal of choice for dissection in many undergraduate courses. The ubiquity of the cat as a pet, cartoon character, and star of television advertisements has provided most students with an opportunity to observe and become familiar with the way that cats sit, stand, walk, run, leap, and perform the other varieties of motion of which they are capable. (The cartoon cats, I must admit, are rather extensively anthropomorphized, so too much reliance cannot be placed on their actions.) This familiarity provides an essential opportunity to compare the cat and the human as study of the skeletal and muscular and other systems proceeds. By drawing attention to the differences between structural features of the two species, a further dimension in understanding function is added.

The organization of this manual may need some explanation. It starts conventionally with laboratory exercises on the cell, tissues, and skeleton, but then goes to a regional approach to the dissection of the cat limb musculature, blood vessels, and nerves, starting with the lower (hind) limb. The reason for this approach is that the hind limb is much simpler in structure and much easier to dissect than the upper (fore) limb. The muscles are larger and fewer, and the innervation is simpler than that of the upper limb. It is, therefore, the better region for the beginner to learn dissection. The same rationale holds true for starting the study of human musculature with the lower limb. Again, the structure is simpler, and this is related to the relatively limited actions of the lower limb of man compared to those of the upper limb. Finally, I have long believed that since the starting point of the evolution of man was the development of bipedal locomotion, the best approach to understanding human structure and function is to start by understanding the significance of bipedal locomotion and the erect trunk.

A second difference in approach concerns the detailed study of the limb skeletal structures that I have taken. Although the fourth laboratory period gives an overview of the skeleton, detailed analysis of the structure of the limb bones is postponed until it can be correlated with the functional significance of various parts as muscle attachments and joint formations with their relation to movement.

The dissection instructions, in most cases, are sufficiently detailed so that some students will be able to work quite independently. Others will receive help from the various photographs of dissections and microscope slides. The photographs are supplemented by line drawings, which in many cases can draw attention to significant detail much more successfully than a photograph.

In the preparation of this manuscript, I have been helped immeasurably by a number of people; Douglas Morton, the publisher, for his encouragement, advice, and remarkable organizational abilities; Dianne Kedro for the cooperative and expert editing; Richard Fritzler and Susan Strawn, who furnished the excellent line drawings from their own dissections; John Bard, one of my students, who was responsible for the time-consuming and meticulous dissections of the shoulder and chest regions. My wife, Jean, deserves special appreciation, not only for typing this and many other manuscripts, but also for acting as a much needed "motivator" when my thoughts strayed toward the tennis courts or the snow-covered hills.

As this is a first edition, in spite of the fact that it has been used for several years in my own course, there will be inconsistencies, inaccuracies, and omissions. For these, I am solely responsible.

LABORATORY 1

Cell Structure

OBJECTIVES

By the end of this laboratory period, you should be able to accomplish the following objectives.

1. Identify the following parts of the microscope on the microscope or a photograph or drawing; and state or recognize a statement of the function of each:

   Substage lamp
   Condenser
   Iris diaphragm
   Stage
   Objectives
   Ocular
   Focusing controls, coarse and fine
   Condenser control

2. Use the microscope rapidly and correctly to locate and focus on specimens for optimum visibility.
3. Prepare a temporary slide of oral epithelium, stain it, and identify the cells, their nuclei, and cytoplasmic granules.
4. Estimate the approximate size of cells in a microscopic field.
5. Prepare temporary stained mounts of fresh tissues such as skeletal muscle, glandular epithelium (liver), lining epithelium (stomach or intestine), and nerve (sciatic nerve).

MATERIALS

1. Microscope
2. Slide of frog red blood cells
3. Microscope slides and cover slips
4. Toothpicks
5. 0.9% NaCl solution
6. 0.5% methylene blue solution (aqueous)
7. Frog tissues

EXERCISE 1
USE AND CARE OF THE MICROSCOPE

The microscope is an expensive and delicate instrument. It is essential to learn to use it properly for both the preservation of the instrument and the speed and ease of making observations. The compound microscope used in biological work achieves its results by magnifying the image of an object as light is passed through it. The light source is below the specimen, and as it passes through the specimen into the objective lens, the image is magnified. The light rays then pass on up the tube to the ocular, where the image is again magnified. The total magnification is the product of the power of the ocular by the power of the objective. A piece of ground glass or copy paper held just above the eyepiece when light is on and a specimen is in position on the stage will show an image of the object that is magnified according to the power of the ocular and the objective. The disc of light formed by this procedure is called Ramsden's disc after one of the early instrument makers upon whom both astronomy and biology depended for their first advances into the extension of the senses (Jesse Ramsden, 1735-1800).

Find the structures listed below, on the microscope you are using. See Figure 1-1.

1. Substage Lamp

The substage lamp is the light source for the microscope. It may be built into the base of the microscope or it may be separate. In the latter case, the lamp may be placed so that its light is reflected through the condenser by a mirror, or it may be placed directly under the condenser after removal of the mirror.

2. Condenser

The condenser is a lens system that focuses the light from the substage lamp so that a solid cone of light is passed upward to the specimen. For optimum results, the condenser must be focused through the use of a field diaphragm in the light source. Since student microscopes do not have this type of light source, the most satisfactory practice for general usage will be to adjust the position of the condenser for the best visual image obtainable with each change of objective or specimen. In some cases where the specimen is not stained, it may be quite difficult to find. The condenser may be dropped to its lowest position in order to increase the contrast and aid in locating the specimen. After the specimen is found, the condenser should be readjusted, as resolution is decreased by the above procedure.

3. Iris Diaphragm

At the base of the condenser is an iris diaphragm that is used to regulate the angle of the cone of light that passes through the condenser to the specimen. The iris diaphragm

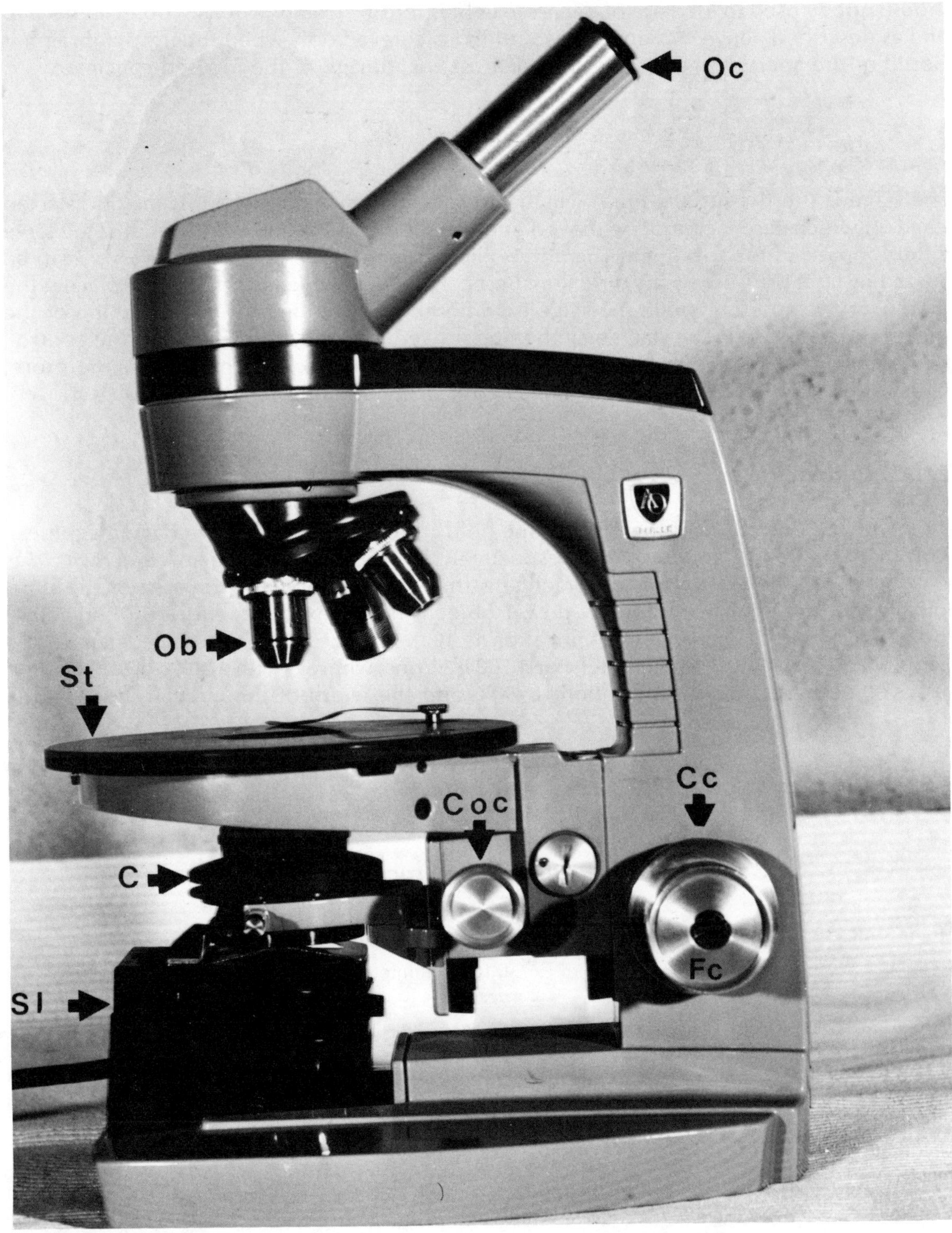

Figure 1-1. The student microscope. Oc—Ocular; Ob—Objective; St—Stage; C—Condenser; Cc—Coarse control knob; Fc—Fine control knob; Coc—Condenser control; SL—Substage lamp.

should not be used to increase or decrease light intensity, except when an object is hard to find as described above. Maximum resolution is achieved only when the iris diaphragm is closed to the aperture which gives the clearest crisp image of the focused specimen.

4. Stage

The stage is the flat surface upon which the specimen slide is placed. It may be a fixed stage, in which case it cannot be moved and the slide must be moved in order to observe different parts of the specimen. The micro-glide stage of AO Microstars, however, can be moved by light pressure in any direction. Best control is achieved in this case by placing the slide over the aperture while the stage is centered and clipping it there with one of the spring clips provided. The stage may then be moved by hand, allowing very fine control. Remember, however, that light rays cross as they pass through the lenses; therefore, objects being viewed will appear to move in the direction opposite to that actually moved.

5. Objectives

The revolving nosepiece of the microscope holds three objectives. These have magnifications of 10X, 40X-43X, and 97X-100X, respectively. *Always start viewing an object with the 10X objective!* The reason for this practice is that the field of view (the area of the slide actually seen) is larger with the low-power objectives. For example, with a 10X objective and a 10X eyepiece, the total magnification is 100X. The width of the circle seen at this magnification is about 1.5 millimeters, or 1,500 microns, whereas with a 40X objective and a 10X eyepiece, the total magnification is 400X, and the width of the circular area seen is

Table 1-1

Comparison of British and Metric Units of Length

Metric Unit	Abbreviation	Metric Equivalent	British Unit
Meter	m	1/1000 kilometer	39.37 inches 1.1 yards
Centimeter	cm	1/1000 meter	0.3937 inches (2.54 cm/in)
Millimeter	mm	1/1000 cm	0.003937 inches (25.4 mm/in)
Micrometer (micron)	—	1/1000 mm	0.00003937 inches (25,400 cm/in)
Nanometer (millimicron)	nm	1/1000	Forget it!

*Most common unit for microscopic measurements of cells.
**Common unit for measurement of cellular organelles and inclusions.

about 0.4mm or 400 microns. Obviously, objects will be located much more rapidly in the larger field. The 97X or 100X, whichever your microscope has, is called the oil immersion objective. This objective is used with the tip immersed in a drop of oil placed over the object being viewed. Do not use it without the oil, as it will give very poor results, and the image you see will not be nearly as clear as with the 40X objective, which is usually referred to as the "high dry" objective.

6. Ocular or Eyepiece

The oculars found in the microscopes used in most undergraduate courses are 10X. Some are wide aperture eyepieces and are convenient for those who wear glasses, since the eye does not have to be so close to the actual lens. The ocular is frequently clouded with fingerprints and should be cleaned with lens tissue before being used. Never use any cloth or paper other than lens tissue to clean lenses. Most papers are hard enough to scratch the soft glass used by lens manufacturers. Lens tissue is a very soft paper made especially for the purpose.

7. Focusing Controls

Locate the large knob on the side of the microscope. This is the coarse focus control and is used to bring the object to rough focus. *Do not use* this control when looking into the eyepiece. In some microscopes, the high-power objectives can be lowered enough to crush the slide, which itself is expensive, but more important, it may cause scratches on the objective — which is very expensive! Watch the position of the objective from the side and, using the coarse control, lower the 10X objective until it stops or until the specimen comes into rough focus. At this point, switch to the fine-control knob (the smaller knob on the side of the microscope), and focus the object sharply. Most modern microscopes are parfocal, i.e., when the object is in focus with the 10X objective, it is possible merely to rotate the nosepiece, bringing the high dry objective into position, and the specimen will be in rough focus. Again, the fine-control knob is used to focus sharply.

8. Summary of Procedure

a. Turn on the substage lamp, and if it is *not* built in, adjust it and the mirror until you see a beam of light coming through the condenser and the stage aperture. Swing the 10X objective into position and observe the pattern of light coming from the condenser. It should be a complete circle with uniform intensity. If not, adjust the mirror and substage lamp until the light is as described above.

b. Place the slide containing the specimen on the stage so that the specimen is directly over the aperture of the stage. Hold it in place with one or two clips.

c. Swing the 10X objective into place, and with the coarse-control knob, lower the nosepiece until it nearly touches the specimen. *Do not look through the ocular to do this; watch from the side.* If your microscope has an automatic stop, it will automatically stop at almost the proper distance for viewing.

d. Look through the ocular focus upward, using the coarse-control knob. Usually the specimen will come rapidly into view. At this point, focus roughly with the coarse control, then switch to the fine control and focus sharply.

e. Adjust the position of the condenser and the iris shutter for maximum viewing and eye comfort.

f. Switch to high dry and repeat step e.

9. Observation of Frog Red Blood Cells

Following the above procedures, observe the permanent slide of frog blood provided.

In the two squares below, sketch in three or four red blood cells in one as you see them under low power and in the other as you see them under high dry. Draw the cells in proportion to the size of the field in each case. Identify the field diameter in each case and from this estimate the size of the cells. Refer to Figures 1-2 and 1-3 for aid in this exercise.

Low Power | **High Power**

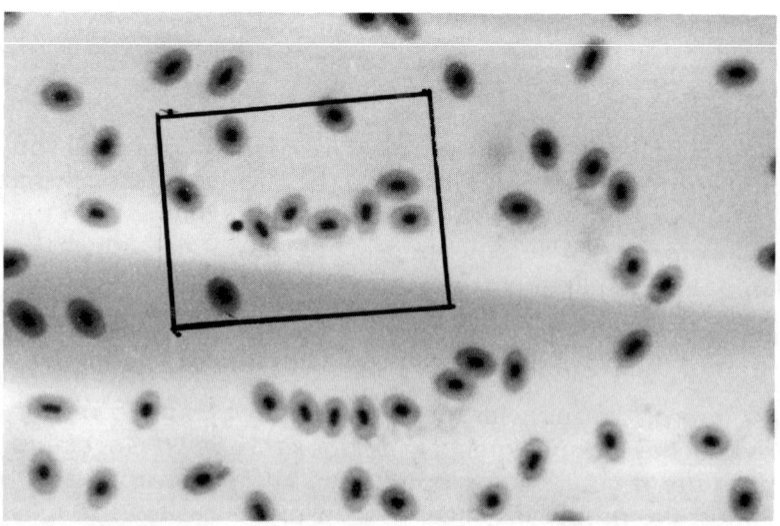

Figure 1-2. Red blood cells of a frog. Low power. The area outlined is shown in Figure 1-3 under high power. Note the reduced size of the field.

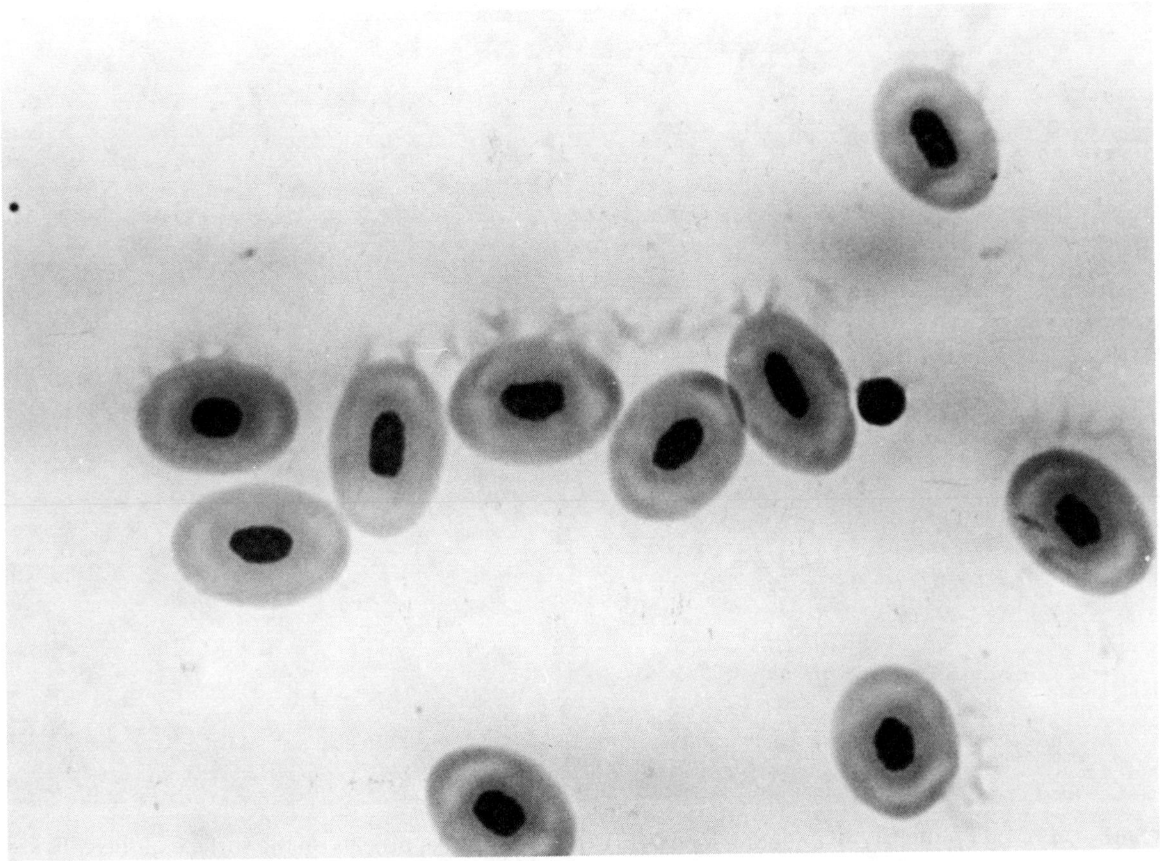

Figure 1-3. Red blood cells of a frog. High power.

EXERCISE 2
PREPARATION OF A TEMPORARY MOUNT OF ORAL EPITHELIUM

The following procedure illustrates a simple method of preparing slides of different tissues for elementary study. More detailed examination of the tissues requires the preparation of a wide variety of permanent mounts, each treated especially to emphasize some particular cellular or tissue feature.

Epithelial Cells from the Lining of the Buccal Cavity

1. Gently scrape the inside of your cheek or lip with a toothpick, and place in a drop of 0.9% saline on a clean microscope slide. Why is 0.9% saline used instead of pure water?

___

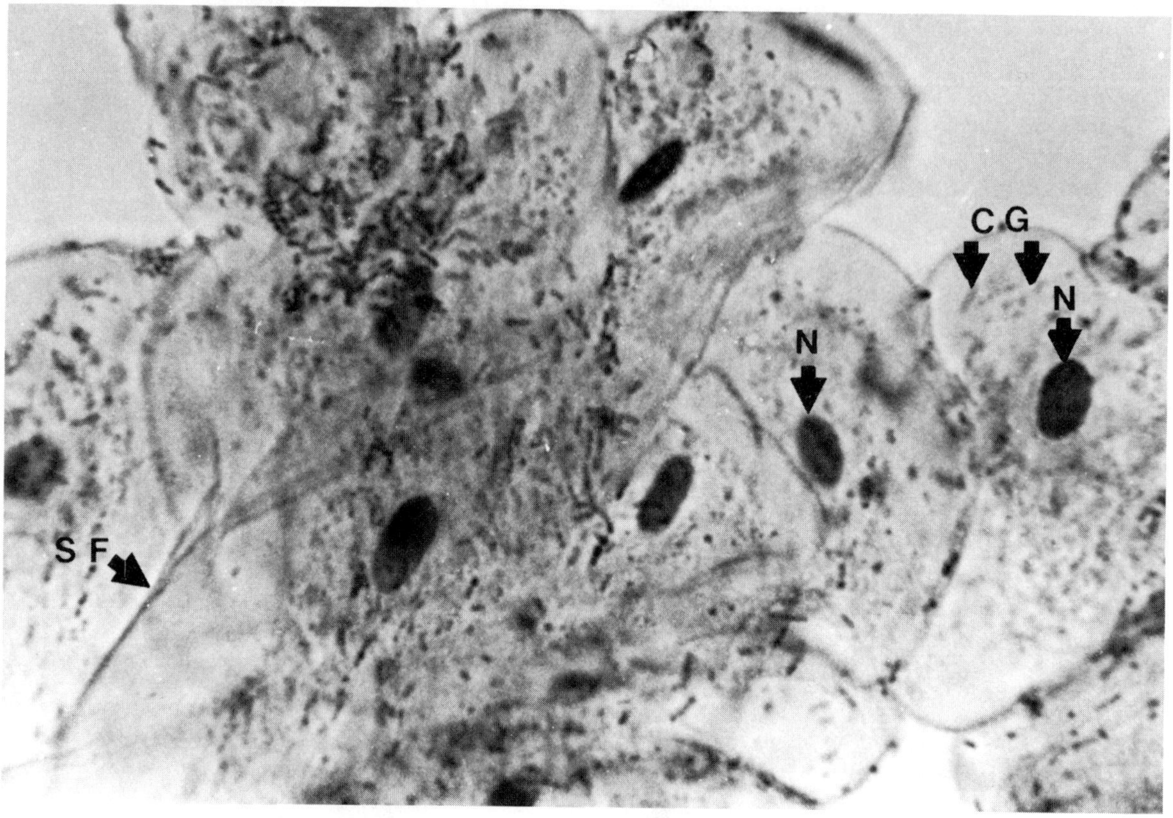

Figure 1-4. Oral epithelial cell under high power. The short, rodlike objects on the surface of the cells are bacteria. N—nucleus; CG—cytoplasmic granules; SF—surface folds.

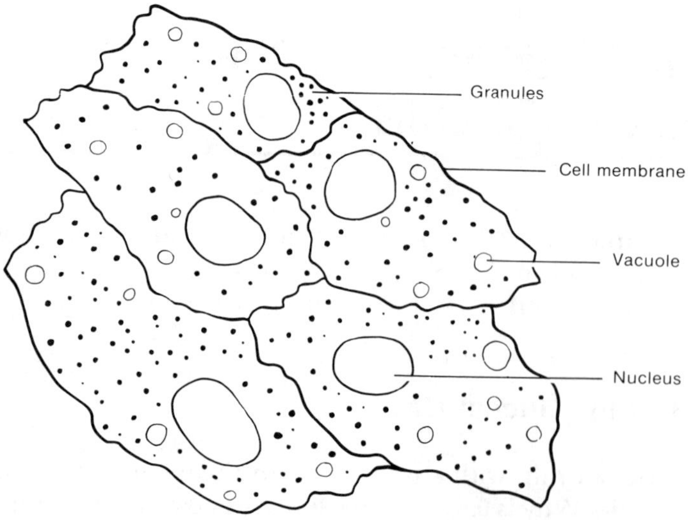

Figure 1-5.

2. Cover the specimen with a cover slip and observe with the microscope under low and high power.

   Can you differentiate cellular detail clearly?

   ____

3. Place a drop of 0.5% methylene blue beside the cover slip, then hold a thin strip of filter paper in contact with the cover slip and saline solution on the side opposite the methylene blue. The filter paper will blot up the saline and draw the methylene blue under the slide in an even-colored layer. Reexamine the cells under the microscope after they have been stained three or four minutes.

   Can you now differentiate more clearly the various structures in the cell?

   ____

   Draw three or four of the cells in the space below. Label cell membrane, cytoplasm, nucleus, granules and vacuoles. Refer to Figure 1-5 to guide you in making your own drawings.

4. What conclusions can you draw regarding the permeability of the cell membrane to the dye?

   ____

   ____

   Which organelle seems to have the greatest affinity for the dye?

   ____

EXERCISE 3
PREPARATION OF TEMPORARY MOUNTS OF LIVER, MUSCLE, AND NERVE CELLS

Cuboidal Epithelium From Frog's Liver

Decerebrate a frog as illustrated by the laboratory instructor. Open the abdomen and remove the liver. Cut a slice from it, and from the cut surface, gently scrape a few cells and stain them with methylene blue as you did previously with the oral epithelial cells. Draw three or four of the cells in the space below, labeling the structural features that can be observed, i.e., nucleus, granules, etc.

Striated Muscle From Frog Leg

Cut a slice of muscle about ½ cm long from the gastrocnemius of the frog. Tease the fibers apart in 0.6% saline and observe them under the microscope. Stain the mount with methylene blue and examine again. Draw a few fibers showing cross striations and nuclei.

Figure 1-6. Skeletal muscle cell. Note nuclei (n) and striations (s).

Figure 1-7. Teased fibers from sciatic nerve of frog. Note myelin sheath (M) and Node of Ranvier (N).

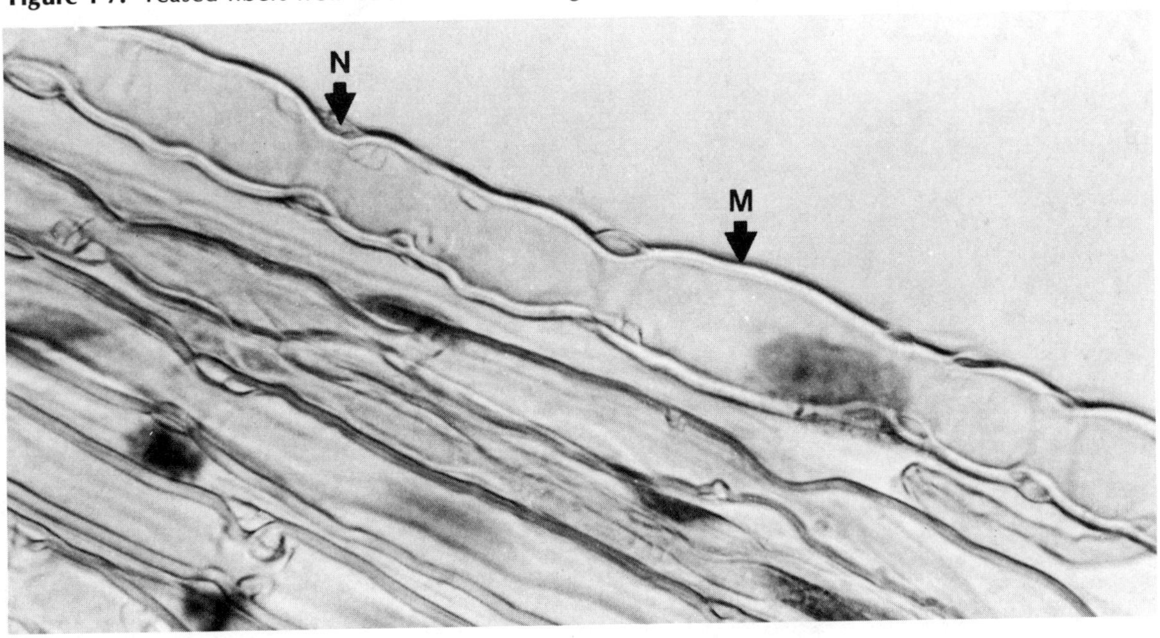

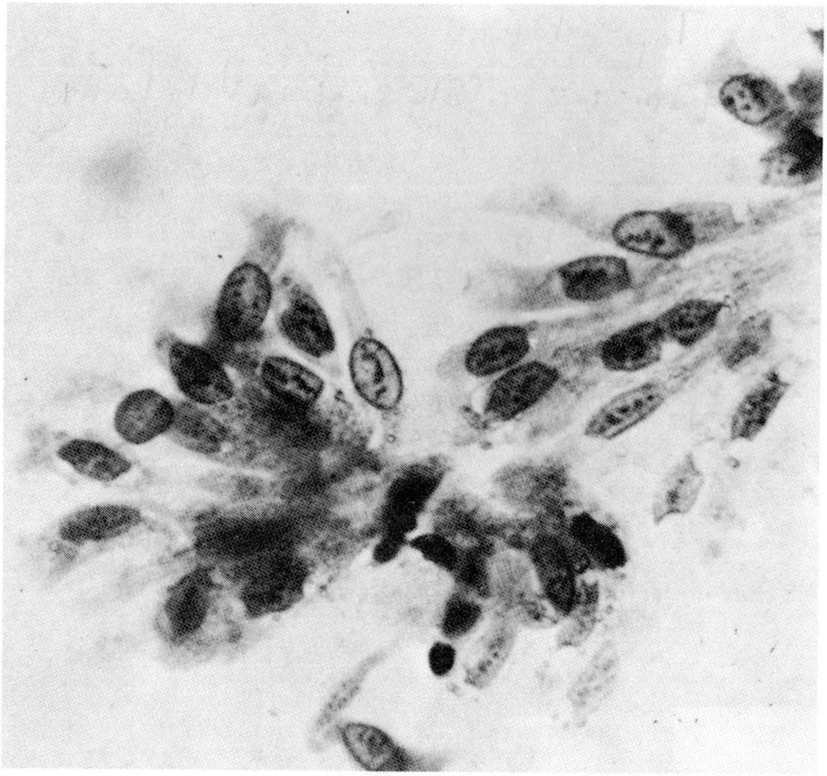

Figure 1-8. Columnar epithelial cells from frog stomach. High power.

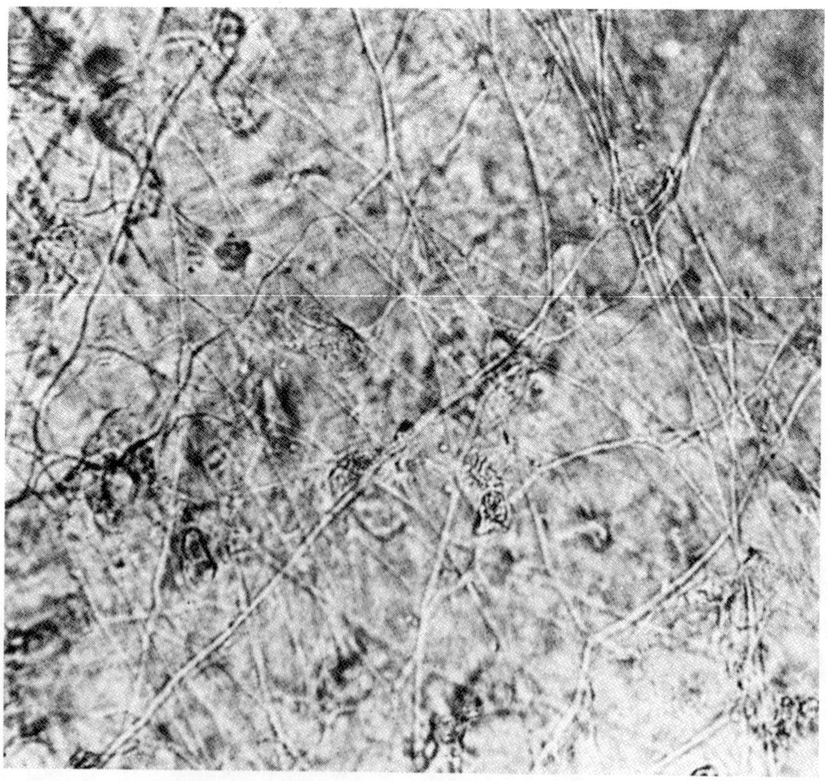

Figure 1-9. Frog mesentery. Broad, wavy fibers are collagenous; narrow, straight fibers are elastic.

Nerve Fibers From the Sciatic Nerve of the Frog

With a pair of dissecting needles, tease out a few fibers from the sciatic nerve of the frog. Mount in saline and cover. Draw methylene blue under the slide and examine under low and high power. Attempt to locate the myelin sheath, nodes of Ranvier, and Schwann cells. Draw a section of fiber showing these features.

LABORATORY 2

Tissues I

Epithelium and Connective Tissues

OBJECTIVES

Upon completion of this study session, you should be able to accomplish the following objectives.

Epithelium

Identify on microscope slides, photomicrographs, or drawings the following examples of epithelium, and state their function, characteristics, and examples of locations where they are found.

1. Simple squamous epithelium
2. Simple cuboidal epithelium
3. Simple columnar epithelium, including absorptive cells and goblet cells
4. Glandular epithelium
5. Pseudo-stratified columnar epithelium
6. Stratified squamous
7. Transitional epithelium

Connective Tissue

Identify on microscope slides, photomicrographs, or drawings the following examples of connective tissue, and state their function, characteristics, and examples of locations where they are found.

1. Areolar connective tissue, including fibroblasts, collagenous and elastic fibers
2 Adipose tissue
3. Dense, fibrous, connective tissue, including fibroblast nuclei and collagenous fibers
4. Hyaline cartilage, including:
   a. Chondrocytes
   b. Lacunae
   c. Matrix
5. Fibrocartilage
6. Elastic cartilage
7. Bone, including:
   a. Osteocytes (if present)
   b. Lacunae
   c. Lamellae
   d. Haversian canal
   e. Canaliculi
   f. Haversian systems or osteons
   g. Interstitial lamellae

MATERIALS

1. Microscope
2. Prepared microscope slides of kidney, intestine, pancreas, hyaline cartilage, and dry bone.

The concept of tissues as cellular aggregates consisting of cells with a common structure and function was first proposed by Marie Francois Xavier Bichat (1770-1802) in 1794, when he was twenty-three years old, and only six years before his death. This brilliant young French anatomist observed that in many widely separated areas of the body there were membranes, clumps of cells, glands, and sheets of cells that were very similar in appearance and that were common to many organs. His observations were based on gross dissection, staining, and maceration techniques. (Maceration involves digesting parts of structures with strong acids, leaving other parts to be observed.) Interestingly enough, he did not employ the microscope. The tissue concept proved to be of enormous importance in understanding organ structure and function both in the normal and diseased organism. Our current, more complete and refined version of the tissue concept has resulted from exhaustive studies of histologists, cytologists, embryologists, and microscopists over the last 175 years since Bichat. It is one of the more impressive attempts of biologists to impose order on Nature. Today we classify tissues into four primary groups: (1) epithelium; (2) connective tissue; (3) muscle; and (4) nerve. Each primary tissue has several subdivisions, and each has certain distinguishing functions and locations.

EXERCISE 1
EPITHELIAL TISSUES

The cells that form the epithelial tissues are found on the surface of the body, lining all body cavities and hollow organs (viscera). In addition, all glands such as the liver or pancreas are classified as epithelial tissues. This is logical, since glands develop during embryonic growth from the lining and covering varieties of epithelium.

Since epithelium occurs in such a wide variety of organs, its cells are responsible for a large number of functions. For example, cells on the surface of the skin protect the underlying softer tissues from mechanical injury, body fluid loss, and other potentially dangerous occurrences. Cells occurring in glands, whether they be multicellular or single cells in the lining of a viscus have a secretory function. Ciliated cells, lining the nasal cavity and trachea, perform a filtration function to prevent small particles from reaching the lungs. Cells in the digestive tract and the kidney have the ability to perform complex transport functions for the absorption of digestive foodstuffs, water, and ions.

Epithelial cells, wherever they might be found in lining surfaces or covering surfaces, generally show the following characteristics.

1. They form sheets or membranes of closely packed cells with little or no intercellular material.
2. The cells forming the sheets or membranes have one free surface and one surface that is attached to the underlying connective tissue by a sticky, fibrous substance called the **basement membrane.**
3. Blood vessels do not penetrate the epithelial membrane. Therefore, the cells must depend upon the exchange of nutrients, respiratory gases, and metabolites with blood vessels lying in the underlying layers of connective tissues.
4. Epithelial tissues have a high regenerative capacity and proliferate rapidly throughout the normal healthy lifespan.

Classification of epithelial tissues is quite simple, depending primarily on cell shape and cell arrangement. The cell shapes are **squamous,** flat, or scaly (an egg sunny-side up is a good model for the shape of a squamous epithelial cell). Cells whose height is approximately equal to their diameter are called **cuboidal**, and elongated cells whose height is greater than their diameter are called **columnar**. The cell arrangement may be **simple**, in which there is only one layer of cells, forming a membrane; or **stratified**, where the cells are several layers thick. Two other descriptive terms, which refer to special cases of epithelium, are **pseudostratified** and **transitional**. In the former case, the cells form a membrane composed of simple columnar epithelium whose nuclei are at different levels, giving the appearance of stratification. **Transitional epithelium** is found in the lining of the urinary bladder, which, when distended, causes the cells to stretch into a stratified, squamous appearance, but in the empty bladder the cells assume a more nearly cuboidal configuration.

In the next few pages, you are instructed to make drawings of a few samples of epithelium as seen in different tissues. Photomicrographs taken from the slides that you will be using are provided to help you find significant features, and line drawings are provided to give you a clearer idea of the three-dimensional appearance of these tissues. Your instructor may wish to have you label these drawings rather than, or in addition to, making your own sketches.

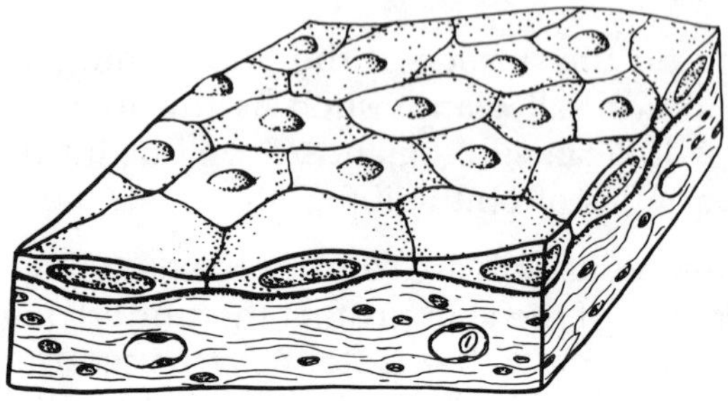

Figure 2-1. Simple squamous epithelium.

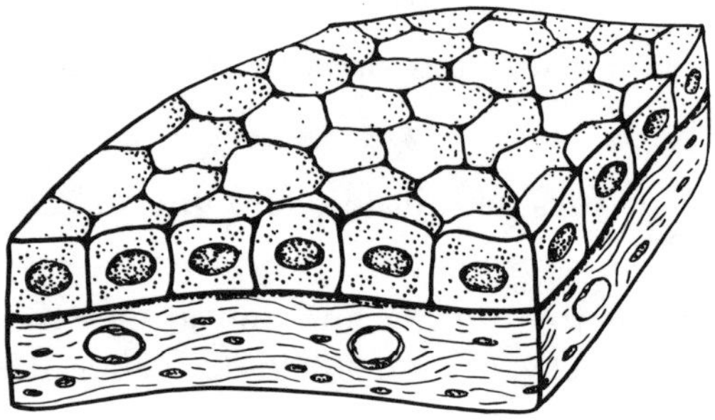

Figure 2-2. Simple cuboid epithelium.

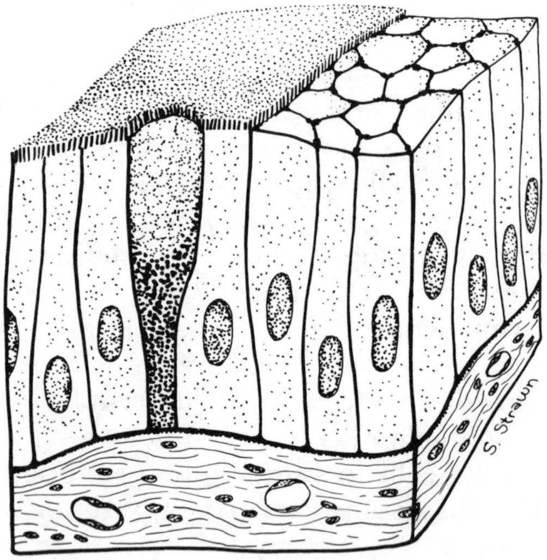

Figure 2-3. Simple columnar epithelium.

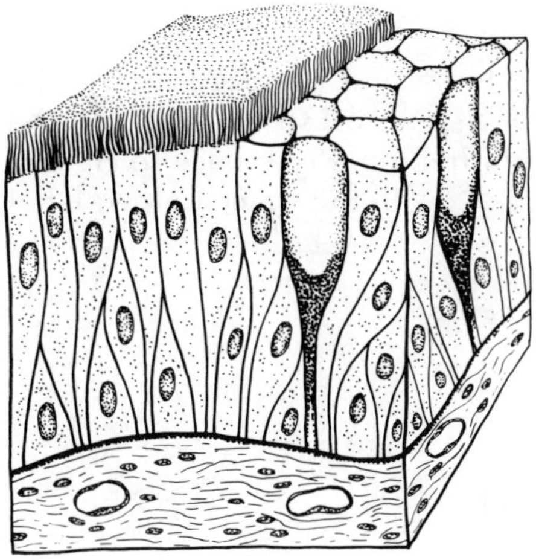

Figure 2-4. Ciliated pseudostratified epithelium.

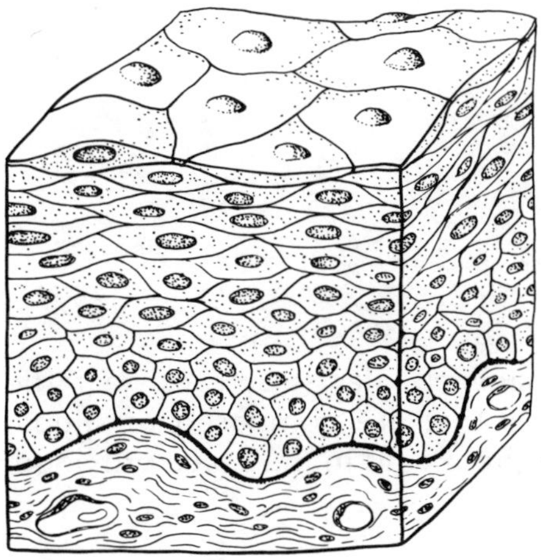

Figure 2-5. Stratified squamous epithelium.

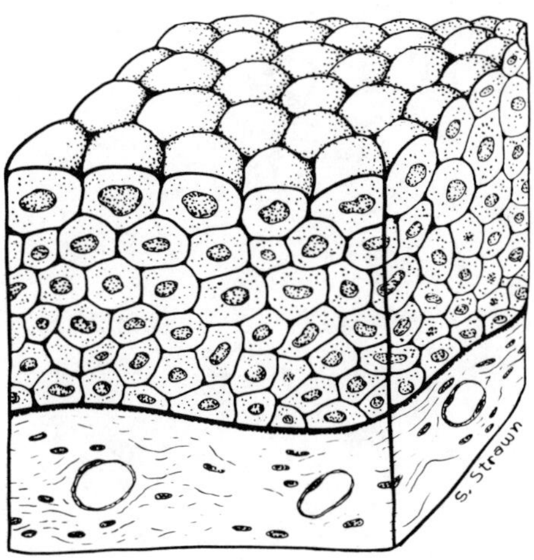

Figure 2-6. Transitional epithelium.

Simple Squamous Epithelium

Place the kidney slide on your microscope and locate a glomerulus (see Figure 2-7). The thin line of cells outlining the glomerulus is the profile of a spherical structure called the Bowman's capsule. The outer layer is simple squamous epithelium. Refer back to Figures 2-1 and 2-5to see the top view of this type of cell. Sketch a few in the space below; label the nucleus, the cytoplasm, and the cell membrane.

Simple Cuboidal Epithelium

Again, using the kidney slide, observe one of the many circular structures near the glomerulus. These are cross sections of kidney tubules that are composed of simple cuboidal epithelium. Refer to Figure 2-7. Draw a few of these cells in the space provided. Label the nucleus, the cytoplasm, and the cell membrane.

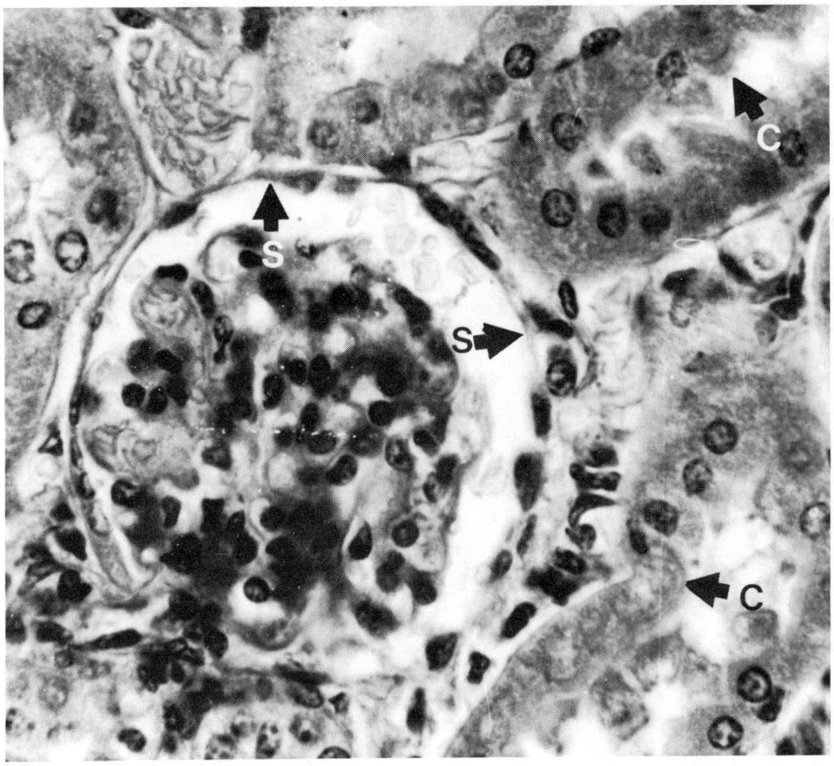

Figure 2-7. Bowman's capsule and glomerulus of kidney. S—squamous epithelium of Bowman's capsule; C—cuboidal epithelium of kidney tubule.

Figure 2-8. Villus of intestine. C—columnar epithelium, absorptive cells; note dark line on free surface—this is the brush border; G—goblet cells, single-celled glands interspersed among the absorptive cells.

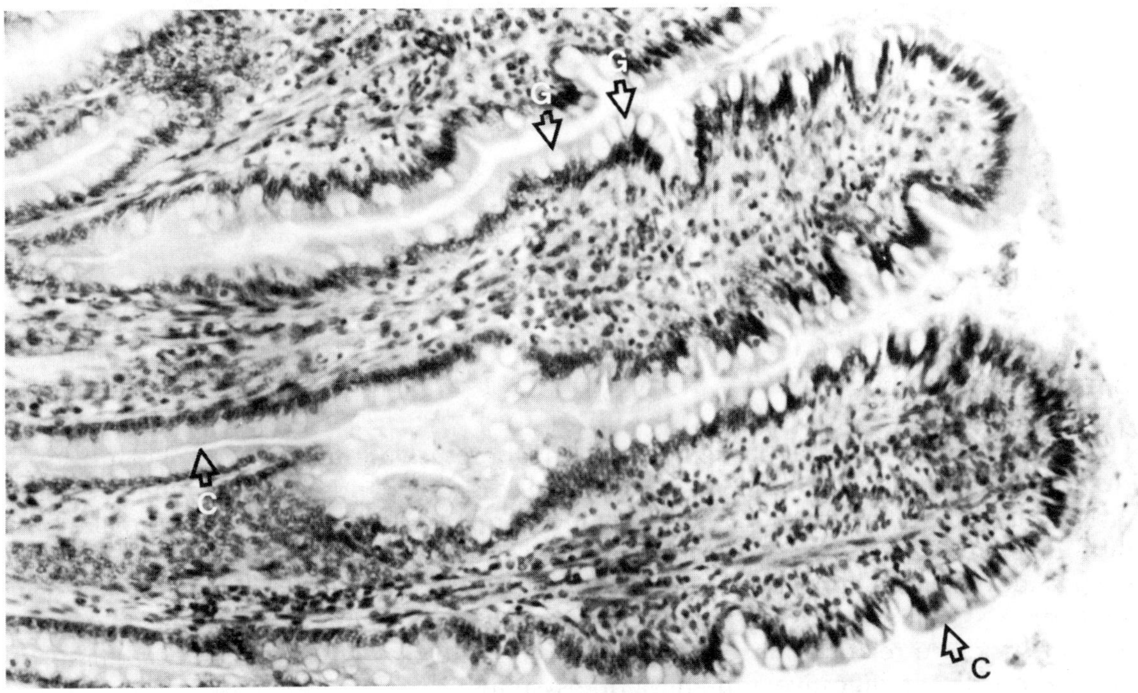

Simple Columnar Epithelium

Remove the kidney from your microscope and replace it with a slide of intestine (may be labeled "Ileum"). Locate one of the many fingerlike processes (villi) protruding into the lumen. Observe the two types of cells that cover the villus. The most common varieties are the tall **absorptive cells**. Interspersed among these are **goblet cells**, easily identifiable by the large, goblet-shaped vacuole in them (see Figure 2-8).

Draw a few of these cells. Label the absorptive cells and the goblet cells.

Glandular Epithelium

It is difficult for the beginning student to get a clear picture of the histology of glands from slides that are usually available. The irregular patterns that glands assume as they develop, and the thickness of a section that generally incorporates several layers of cells make for a confusing picture. Over the years, histologists and embryologists have, by using a variety of methods, developed a concept of the growth of exocrine and endocrine glands and have settled on a system of classification. These are shown in figure 2-9 on the next page. Function-structure relationships will become more clear as we study specific glands in connection with the digestive and endocrine systems.

Figure 2-9. Types of exocrine glands.

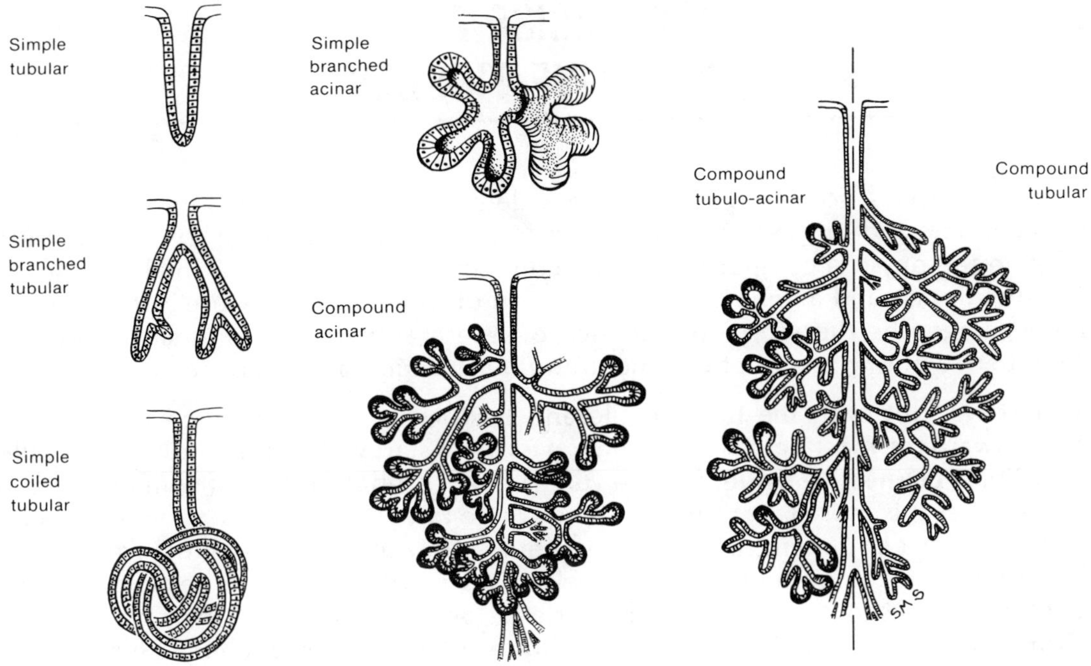

Figure 2-10. Development of glands.

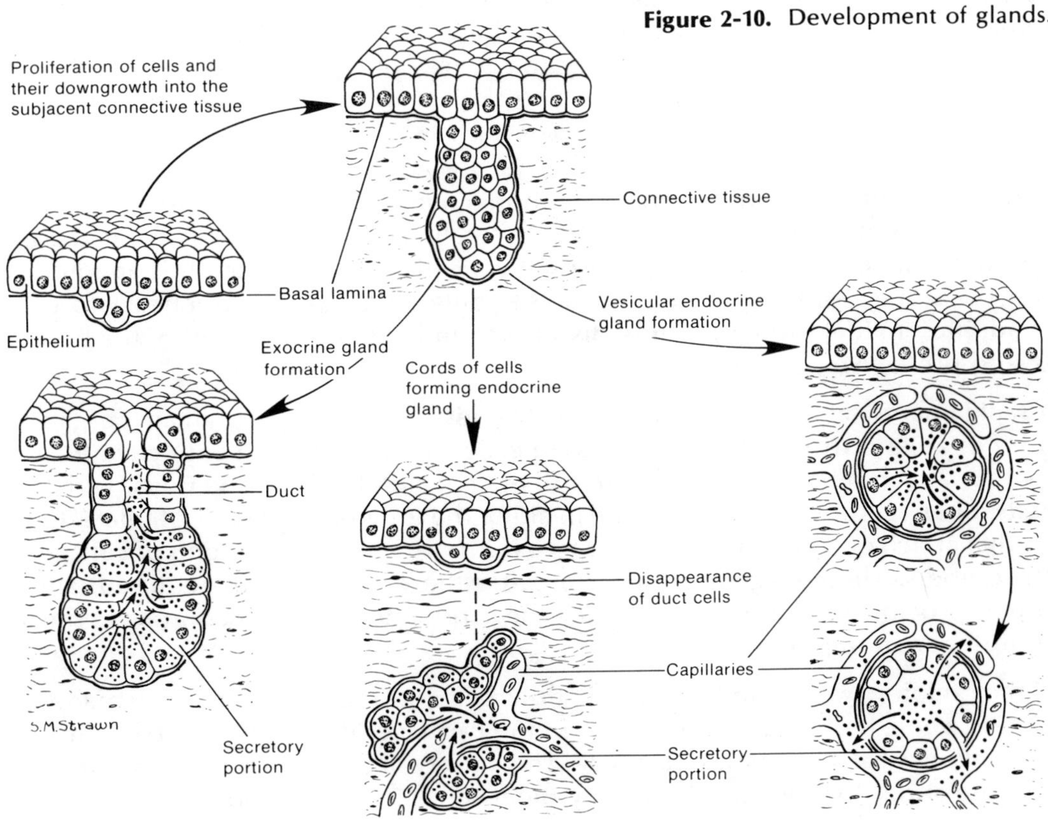

EXERCISE 2
CONNECTIVE TISSUES

As the name implies, connective tissues are widely distributed throughout the body and serve to connect and support other softer tissues. There are many subvarieties and much variation in the physical composition of the connective tissues. This fact makes classification a difficult task and a generalized description of function even more difficult. In the next few paragraphs, therefore, the function of several subtypes of connective tissue will be described, along with a brief comment on their location and characteristics.

If we start working with the most fragile and flexible forms of connective tissues and proceed toward the most rigid and toughest, we would start with **areolar**, sometimes called **loose ordinary connective tissue**. This is a very widely distributed form of connective tissue since it attaches the entire inner surface of the skin to underlying tissues, forms connecting membranes between organs, and lines the entire thoracic and abdominal cavities. It is made up of three components — a semigelatinous matrix, two or three types of fibers depending upon its location, and several varieties of cells. The term "matrix" refers to a nonliving secretion containing the cells and fibers of most forms of connective tissues. It varies in consistency from a jellylike substance to the very hard material of bone and enamel. The two most common varieties of fibers in areolar connective tissue are **collagenous** or **inelastic** fibers, and **elastic** fibers. In some locations, such as glands, small **collagenous** fibers form a complex network and are called **reticular**. You will not study **areolar connective tissue** in the laboratory this week, but you will be working with it extensively at the gross anatomical level as you skin and dissect various structures in the cat.

A second rather soft and flexible variety of connective tissue is **adipose tissue**, or fat. In contrast to the areolar, **adipose tissue** is nearly totally made up of closely packed cells, called **adipocytes**, held together by a loose network of reticular fibers. It is an important depot for the storage of lipids, a fuel substance containing a very high energy to weight ratio. In addition to fuel storage, **adipose tissue** is found in many locations of the body as a packing and protective padding tissue.

Tendons, ligaments, and the tough, connective tissue coverings of many organs are composed of a very strong and flexible tissue called **dense fibrous connective tissue**. In tendons that attach muscles to bones, the fibers are nearly all **collagenous**, with a few **fibroblasts** (fiber-secreting cells) squeezed in along the long parallel rows of fibers. In joint capsules that attach one bone to another, there is a more equal distribution of **collagenous** and **elastic** fibers. The **collagenous** fibers provide the great strength required, and the **elastic** fibers provide for elasticity, permitting the capsule to stretch or contract as the joint changes in shape during movement.

Most of the skeleton, with its important functions of support and protection, is made up of **cartilage** and **bone**. In the adult body, most of the cartilage has a glassy matrix with widely dispersed cells in small chambers or **lacunae**. This variety of cartilage is called **hyaline** and is found on the surface of bones where they meet in a joint. Its smooth, slippery surface

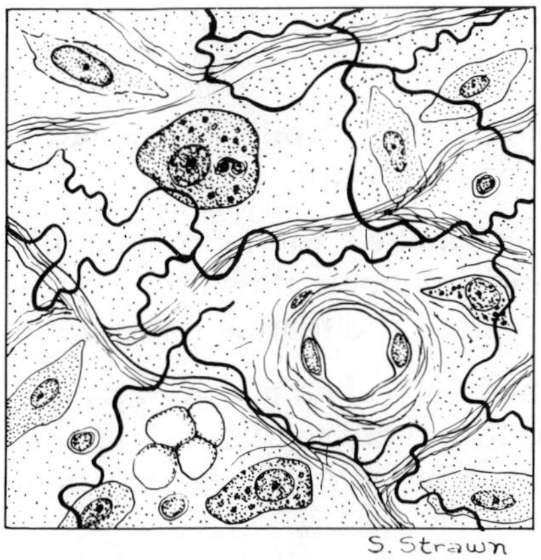

Figure 2-11. Areolar connective tissue.

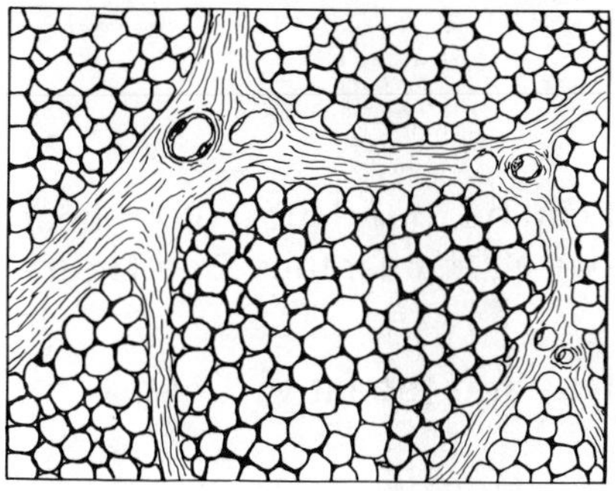

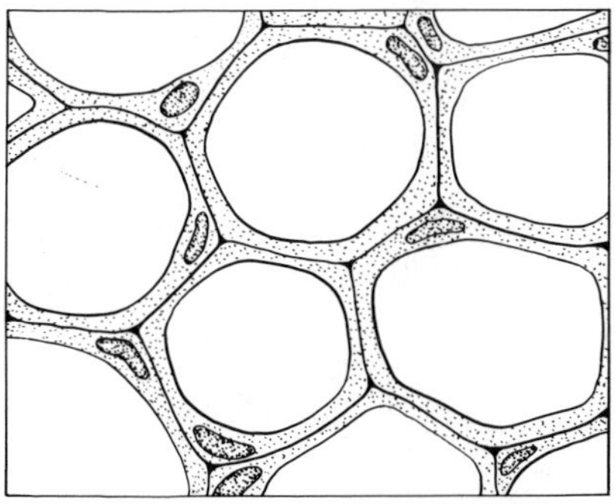

Figure 2-12. Adipose tissue.

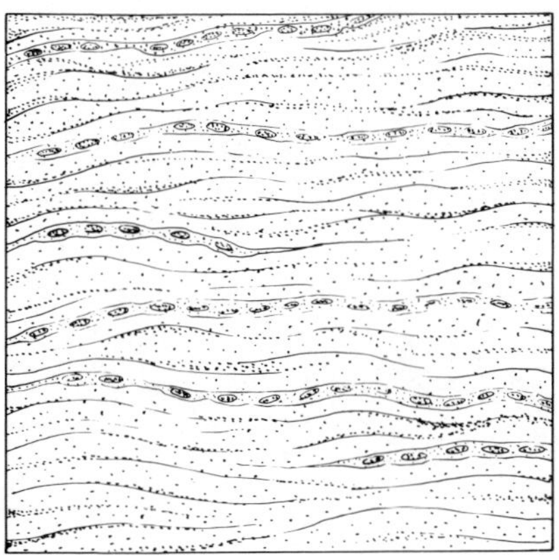

Figure 2-13. Dense fibrous connective tissue.

Figure 2-14. Hyalin cartilage.

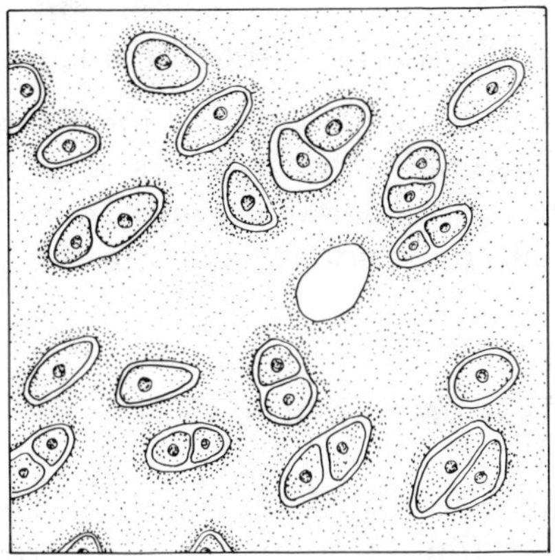

Figure 2-15. Fibrocartilage.

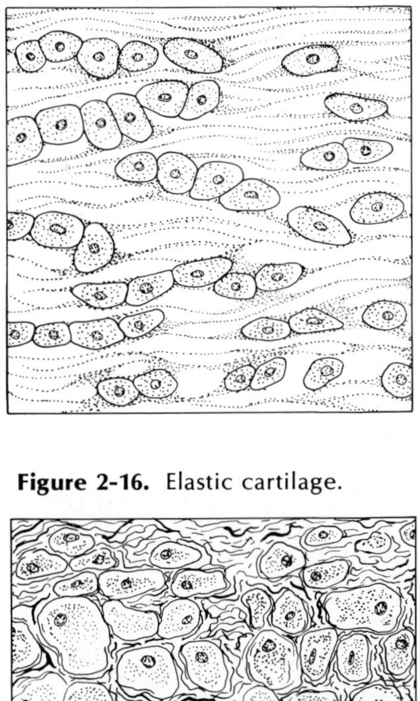

Figure 2-16. Elastic cartilage.

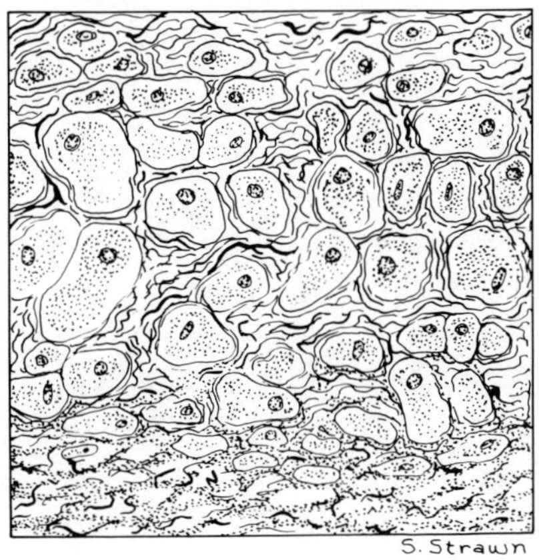

makes it an ideal bearing surface for the reduction of friction as bones move against each other.

In addition to being found in the surfaces of bones forming a joint, **hyaline cartilage** is found joining ribs to the sternum and reinforcing the trachea and bronchi.

Two other varieties of cartilage are **elastic** and **fibrous. Elastic cartilage** is found in the pinna of the ear and the epiglottis, while **fibrocartilage** is found between the discs of the vertebrae and within the knee, where a great resistance to twisting and tearing is required.

Bone is composed of an inorganic matrix of a calcium phosphate salt called **hydroxyapatite.** These hard crystals are reinforced by collagenous fibers and form very regular cylindrical structures called **osteons.** Your instructor may wish to have you label these drawings rather than, or in addition to, making your own sketches.

Adipose Tissue

Return to the kidney slide and locate the adipose tissue that can be found most often near the **hilum** (indentation). It is recognizable by its "chicken wire" appearance (see Figure 2-12 and 2-19).

Draw a few cells below. Label the cell membrane, the nucleus and the fat vacuole.

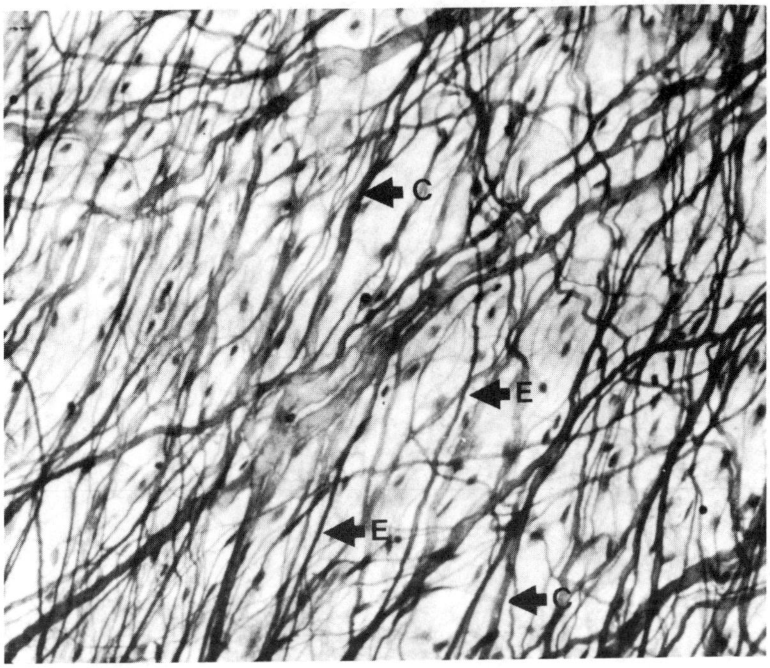

Figure 2-17. Low-power photograph of loose areolar connective tissue. C—collagenous fibers; E—elastic fibers.

Figure 2-18. High-power photo of same tissue. Note variety of cell types in this photograph.

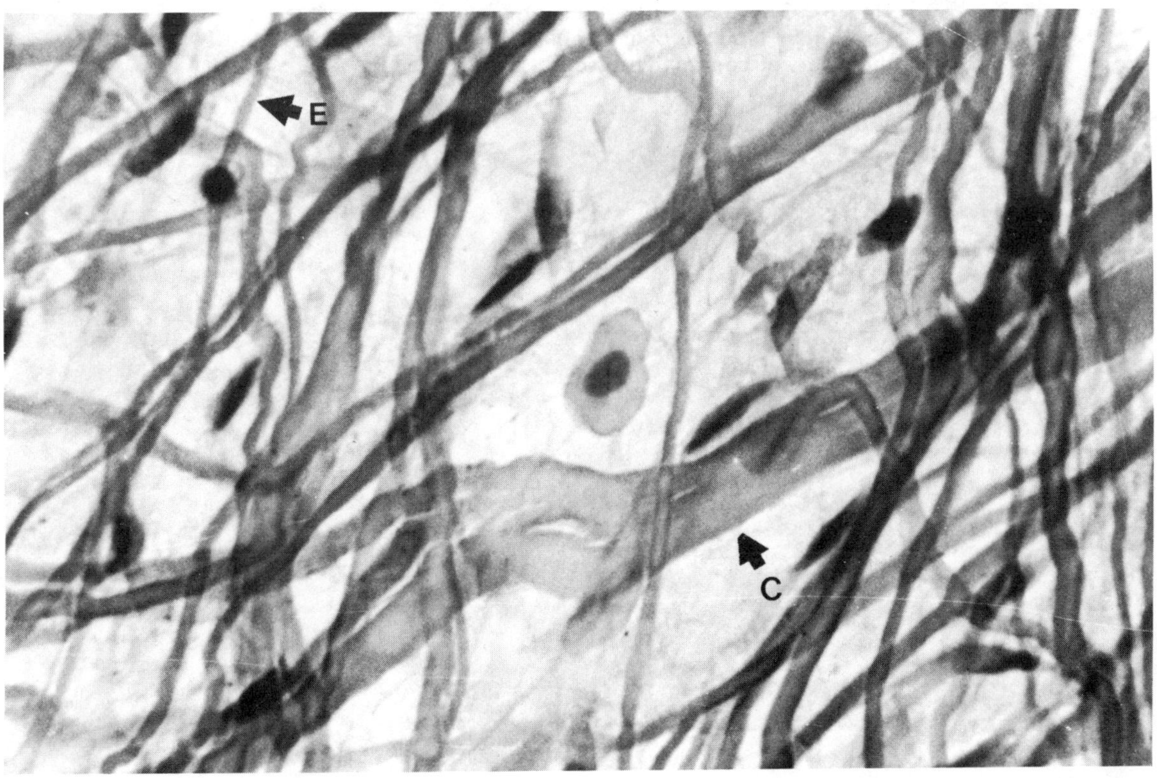

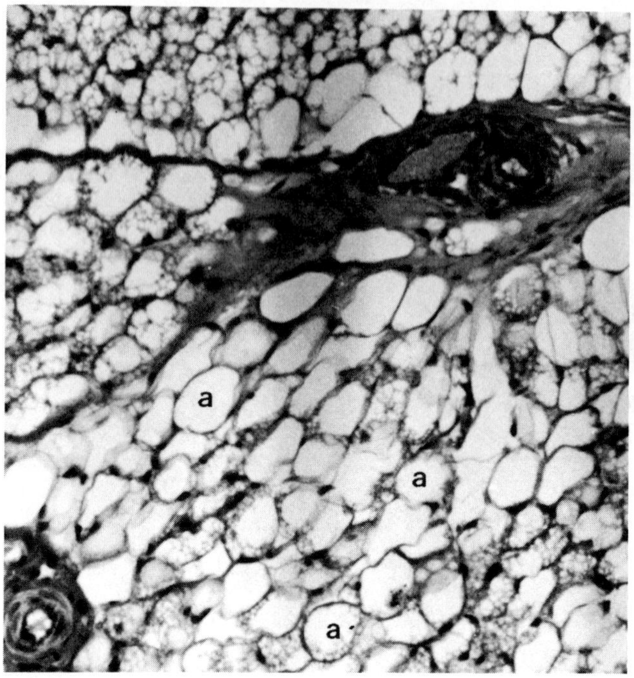

Figure 2-19. Adipose tissue. A—Adipose cells, consisting of a large vacuole and a thin rim of cytoplasm.

Figure 2-20. Dense fibrous connective tissue, from a tendon. F—Nucleus of a fibroblast, the cell that secretes the fibers of connective tissue.

Dense Fibrous Connective Tissue

Observe the demonstration slide of tendon at the demonstration table. Note the abundance of fibers and the nuclei of the fibroblasts squeezed in between the fibers.

Hyalin Cartilage

Place the hyalin cartilage slide on your microscope and locate the following structures.

1. Matrix
2. Lacuna
3. Chondrocyte

Draw a few cells and label the structures listed above.

Bone

Place a dry, ground bone slide on your microscope. Locate an osteon and observe the following structures.

1. Lacunae
2. Lamellae
3. Haversian canal
4. Canaliculi
5. Interstitial lamellae

Draw an osteon, and label the structures listed above. Label these same structures on Figure 2-22.

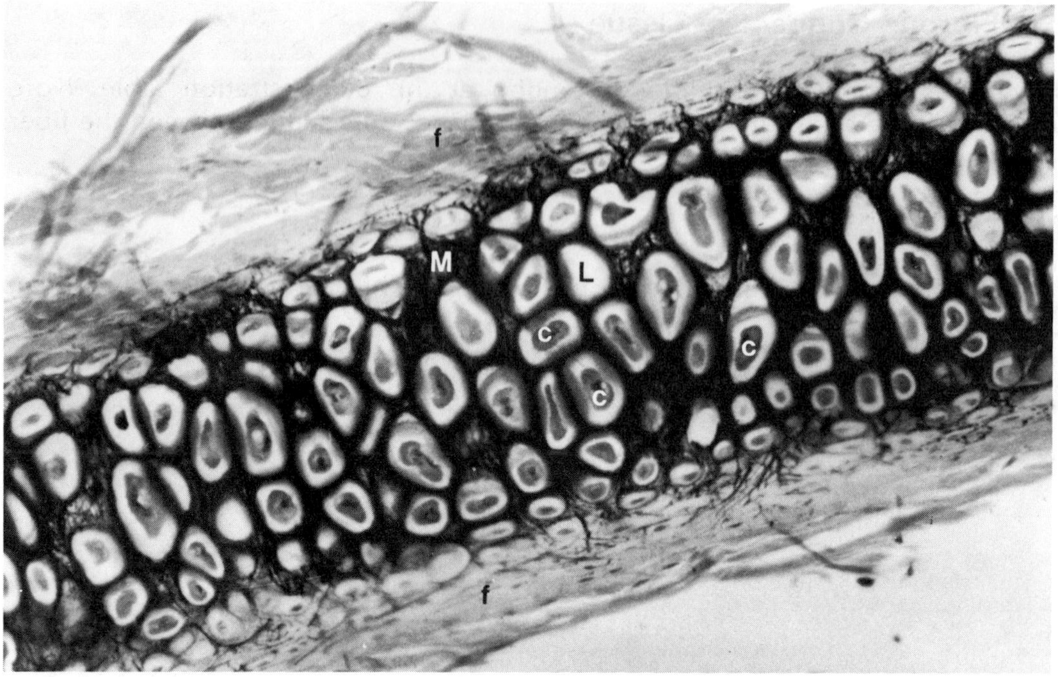

Figure 2-21. Hyaline cartilage. L—lacuna; M—matrix; c—chondrocyte; F—fibrous connective tissue.

Figure 2-22. Dry, ground bone showing osteons and interstitial lamellae. H—Haversian canal; L—lacuna; c—canaliculi.

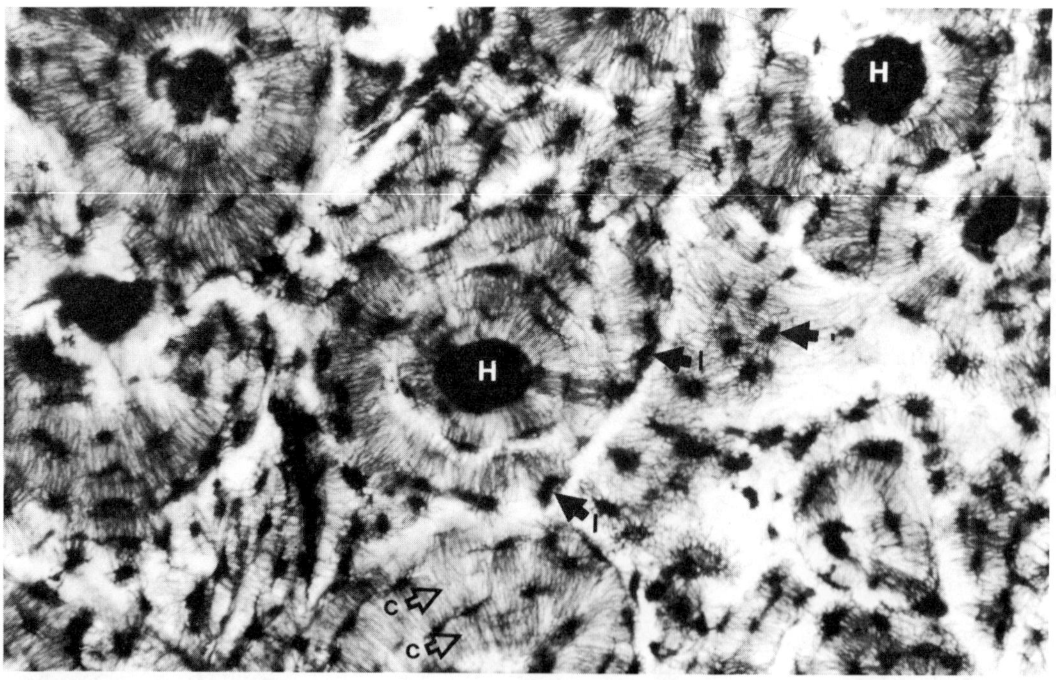

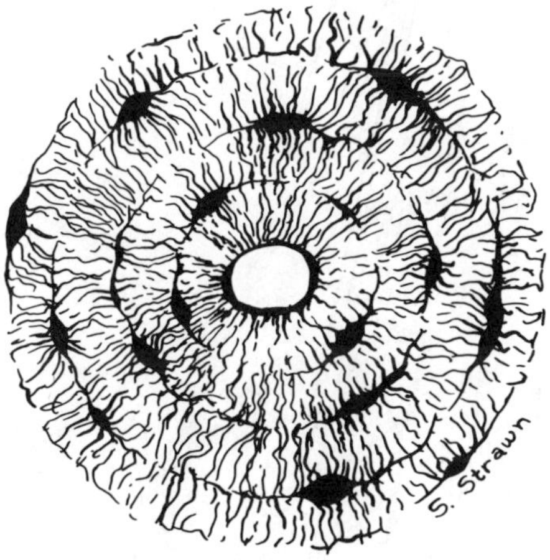

Figure 2-23. Cross section of bone under high power.

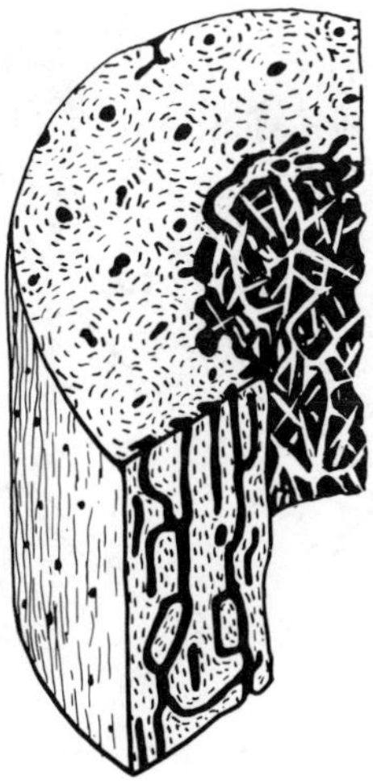

Figure 2-24. 3-D drawing of bone under low power.

Figure 2-25. Schematic drawing of an Haversian system.

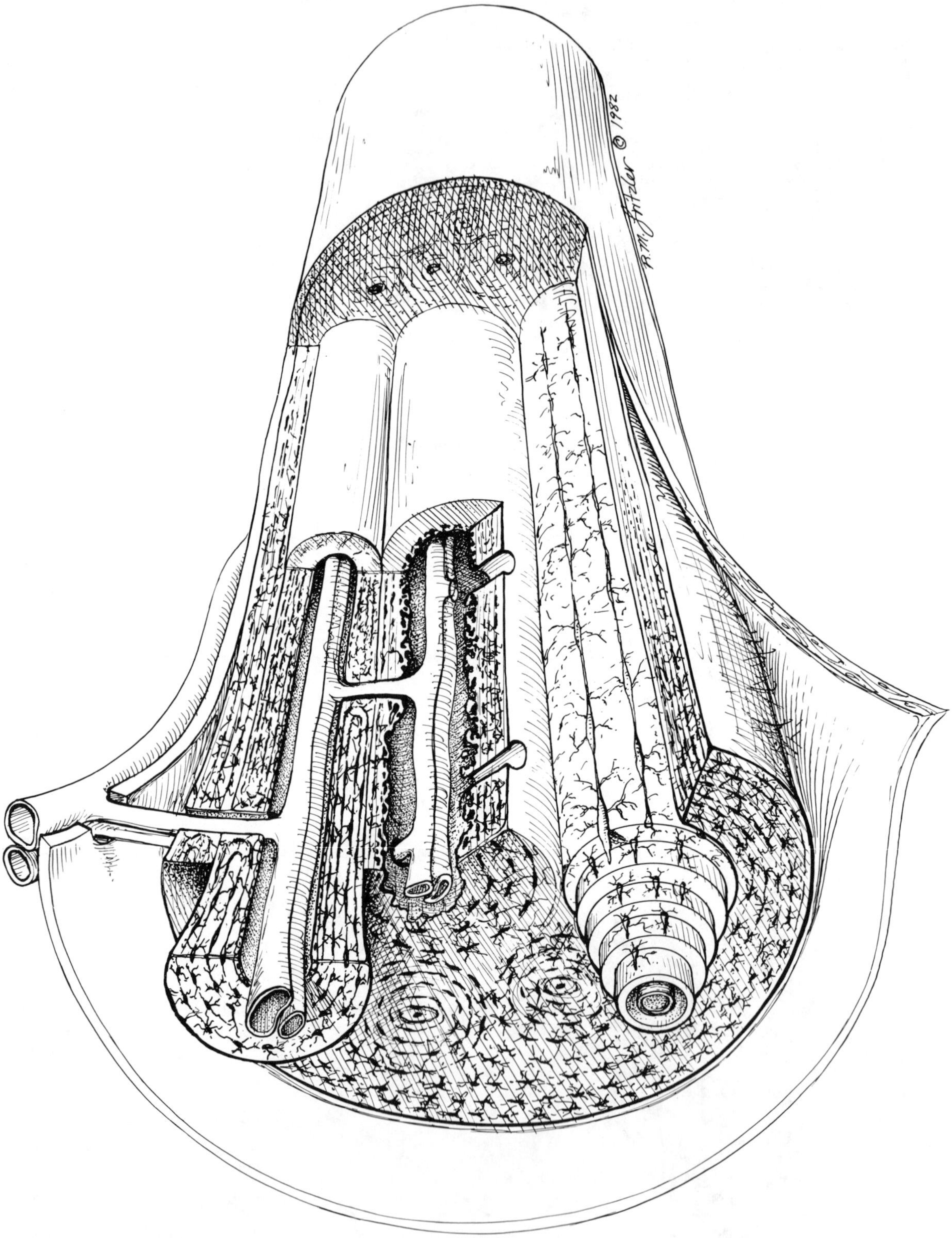

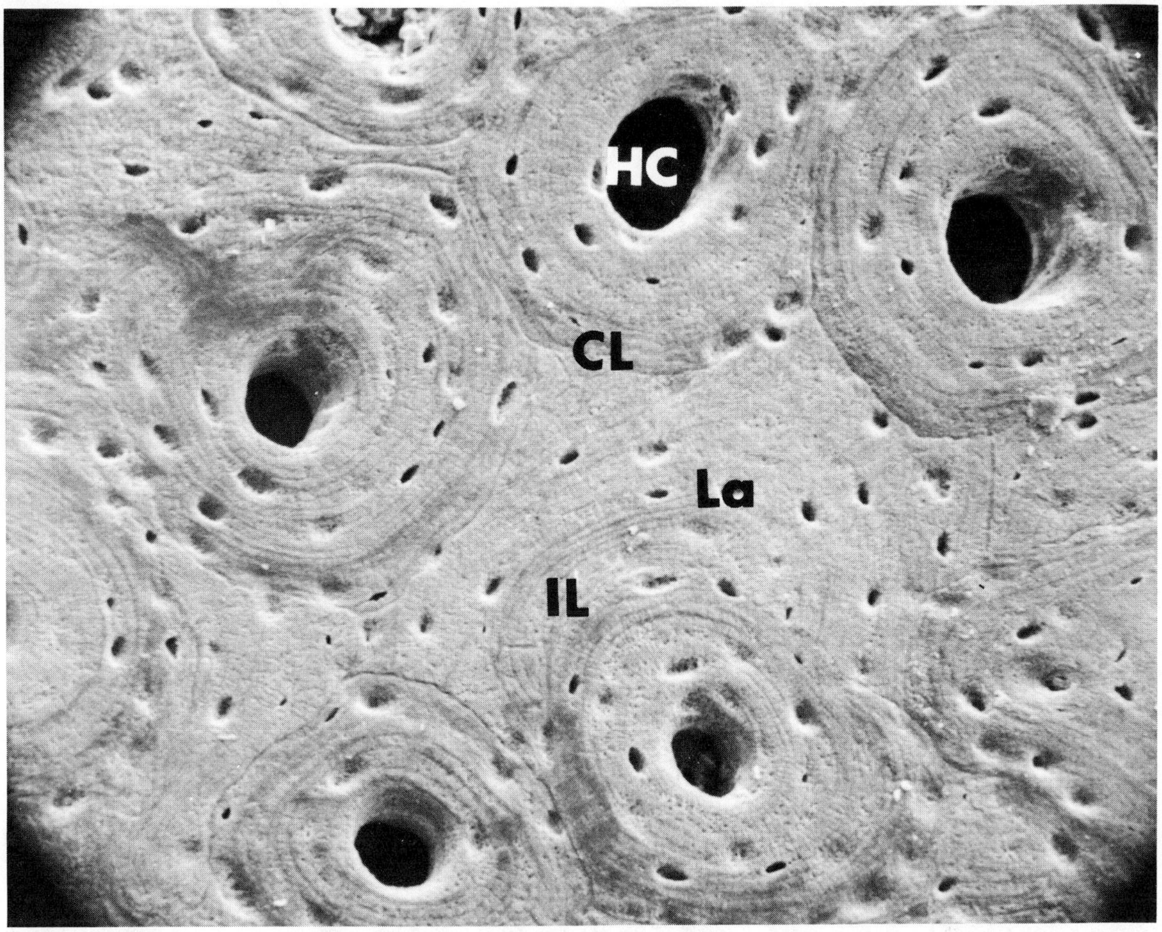

Figure 2-26. Scanning electron micrograph of dry ground bone, 490X. HC—Haversian Canal; CL—cement line; La—lacunae; Il—interstitial lamellae.

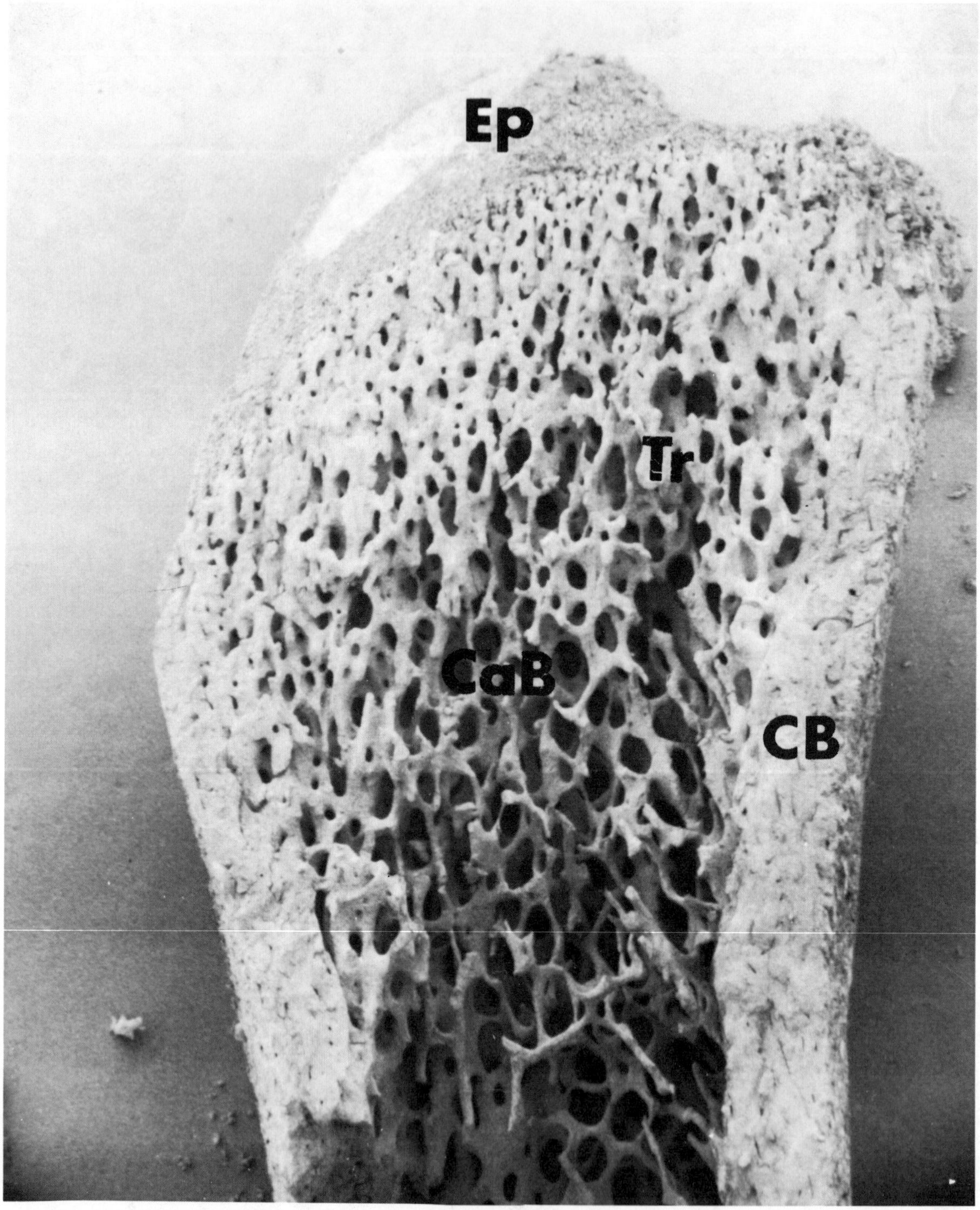

Figure 2-27. Scanning electron micrograph of bone, 40X. Ep—epiphyseal plate; Tr—trabeculae; CaB—cancellous bone; Cb—compact bone.

Figure 2-28. Scanning electron micrograph of dense fibrous connective tissue, 1840X. Cf—collagenous fibers.

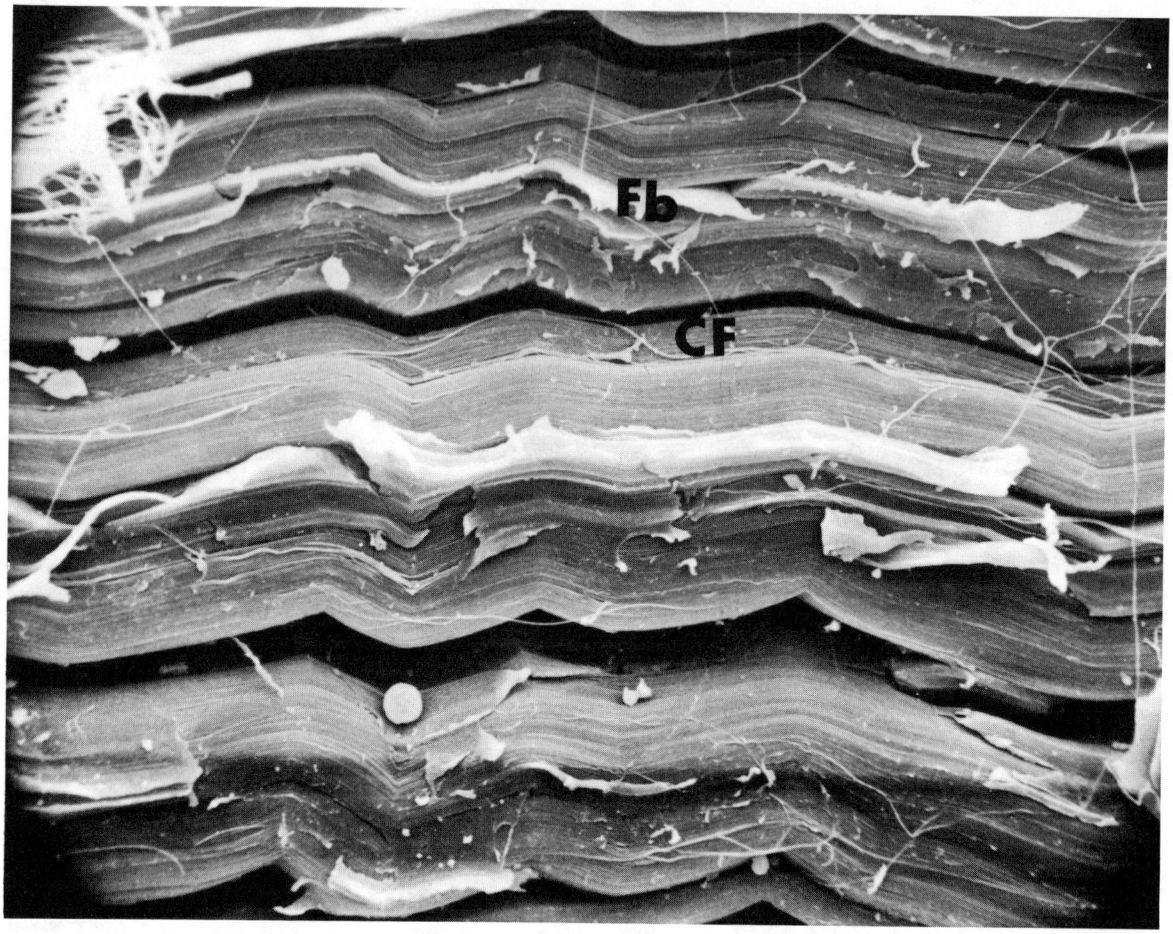

Figure 2-29. Scanning electron micrograph of dense fibrous connective tissue, 690X. Fb—fibroblast.

LABORATORY 3

Tissues II

Muscle and Nerve

OBJECTIVES

After finishing this study session, you should be able to identify the following structures on drawings, photographs, or microscope slides.

Cross Section of Mammalian Skeletal Muscle

1. Cell membrane (Sarcolemma)
2. Nuclei
3. Nerves
4. Arteries
5. Veins
6. Capillaries
7. Epimysium
8. Perimysium
9. Endomysium

Longitudinal Section of Mammalian Skeletal Muscle

1. Striations
2. Nuclei
3. Myofibrils
4. Capillaries
5. Endomysium
6. Perimysium
7. Epimysium

Cross Section of the Spinal Cord

1. Cell body
2. Nucleus
3. Dendrites
4. Axon

Teased Nerve Fiber

1. Myelin sheath
2. Nodes of Ranvier
3. Internode
4. Schwann cells

EXERCISE 1
SKELETAL MUSCLE

The function of muscle, wherever it may be found in the body, is to cause movement by contracting. Skeletal muscle causes movement of joints in various forms of limb, head, and trunk activity. Visceral muscle moves fluids through the tubes of the digestive tract and regulates the diameter of blood vessels to shunt blood from one area of the body to another. Cardiac muscle forms the walls of the heart and propels blood through the vessels of the circulatory system by the force of its contraction. These three varieties of muscle are often described in different terms. Table 3-1 shows the relationships of three different types of classification.

Table 3-1

Descriptive Terms Applied to the Three Classes of Muscle

Classified by Location	Classified by Microscopic Appearance	Classified by Neural Control
Skeletal	Striated	Voluntary
Visceral	Smooth	Involuntary
Cardiac	———	Involuntary

Cross Section of Muscle

Place the cross section on your microscope and locate the following structures:

1. Sarcolemma
2. Nuclei
3. Nerves
4. Arteries and veins
5. Capillaries
6. Epimysium
7. Perimysium
8. Endomysium

Draw a portion of the muscle slide showing the above structures.

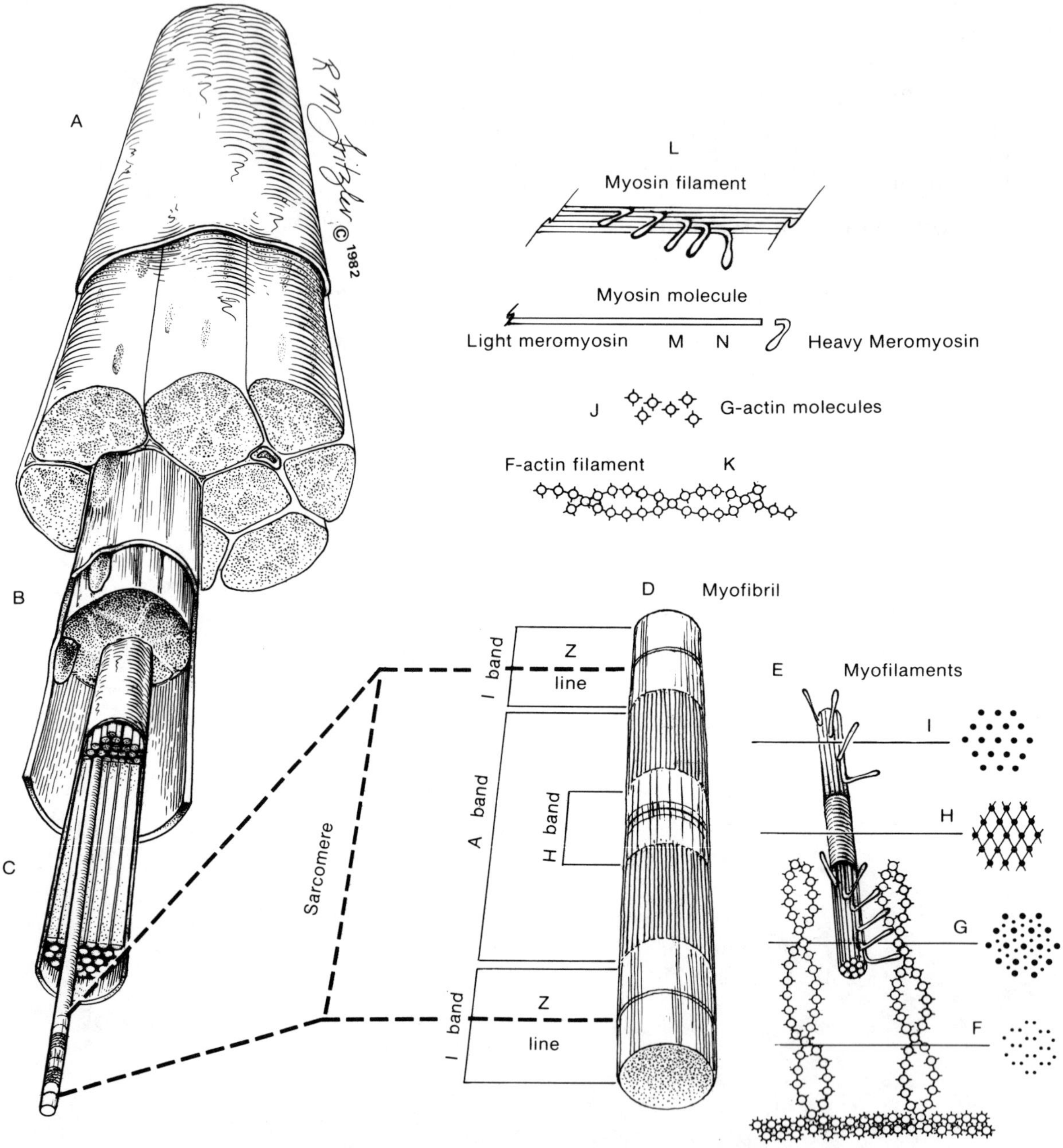

Figure 3-1. Muscle subdivided into successively smaller elements. A—entire muscle; B—muscle cell or fibers; C—myofibrils.

Figure 3-2. Schematic drawing of a muscle fiber showing myofibrils with associated sarcoplasmic reticulum, T-system, and sarcomeres. Label these.

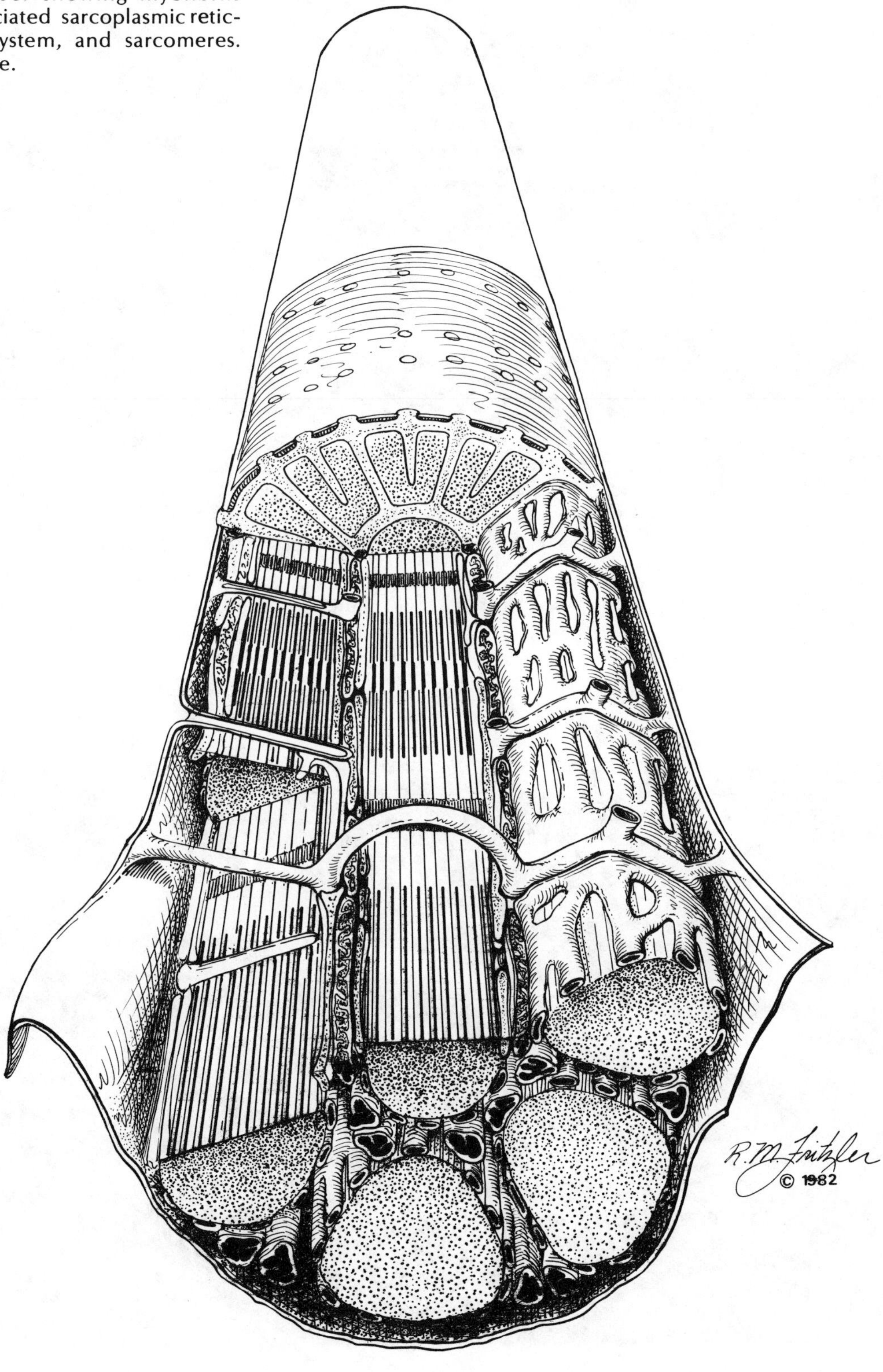

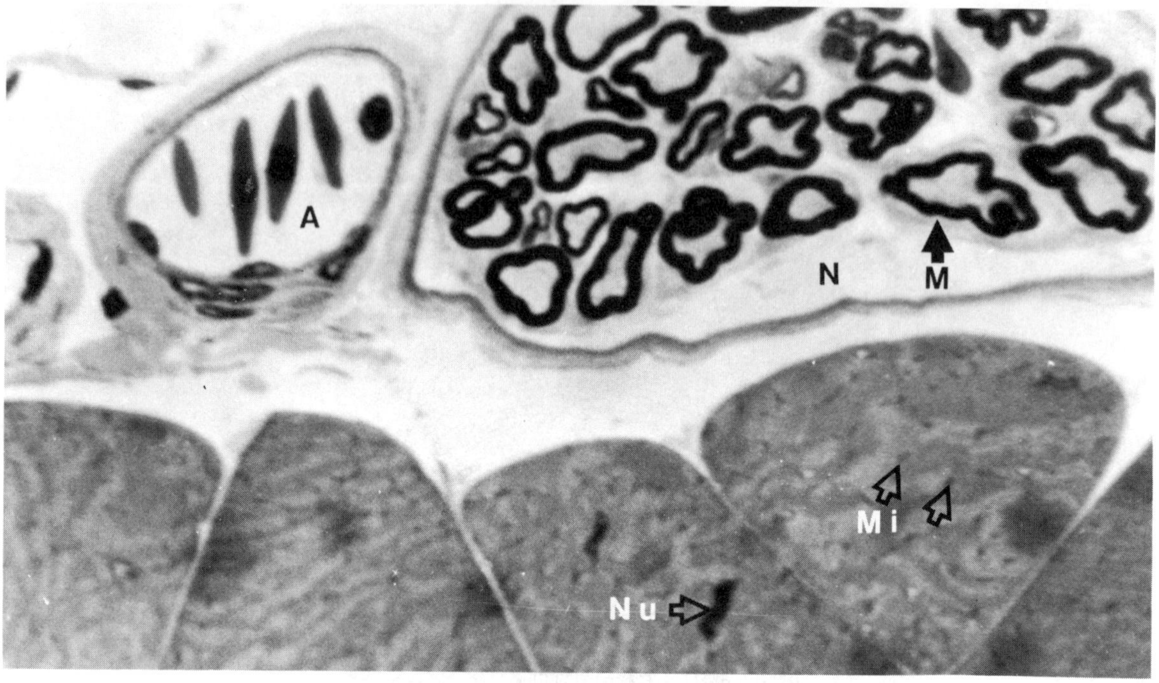

Figure 3-3. Cross section of frog muscle showing portions of 5 fibers. M—myelin sheath of neuron; A—artery with several red blood cells; N—nerve; Nu—nucleus of muscle cell; Mi—mitochondria, high power.

Figure 3-4. High-power photograph of a longitudinal section of frog muscle. S—striations; E—fragments of endomysium; N—nucleus.

Longitudinal Section of Muscle

Place the longitudinal section on your microscope and locate the following structures:

1. Striations
2. Nuclei
3. Myofibrils
4. Capillaries
5. Epimysium
6. Perimysium
7. Endomysium

Draw two or three fibers showing the structures listed above. Refer to Figures 3-3 and 3-4 for help in completing this assignment.

EXERCISE 2
NERVE TISSUE

Nerve tissue is located in the brain and spinal cord as well as in the nerves that emerge from these centrally located structures. The neuron, which is the unit of structure and function of the nervous system, appears in a wide variety of forms. In nearly all cases, however, we find a cell body with many multibranching processes. These processes form the physical basis of the vast communication system represented by the nervous system. Estimates of the number of neurons in the human brain vary from ten to one hundred billion, and since each neuron may be in contact with as many as one thousand others, we can see why the complexity of such a system has baffled human understanding for so long.

Some of the important structures in the system can be observed in slides of the spinal cord and of a teased nerve preparation.

Figure 3-5. Schematic drawing of a motor neuron.

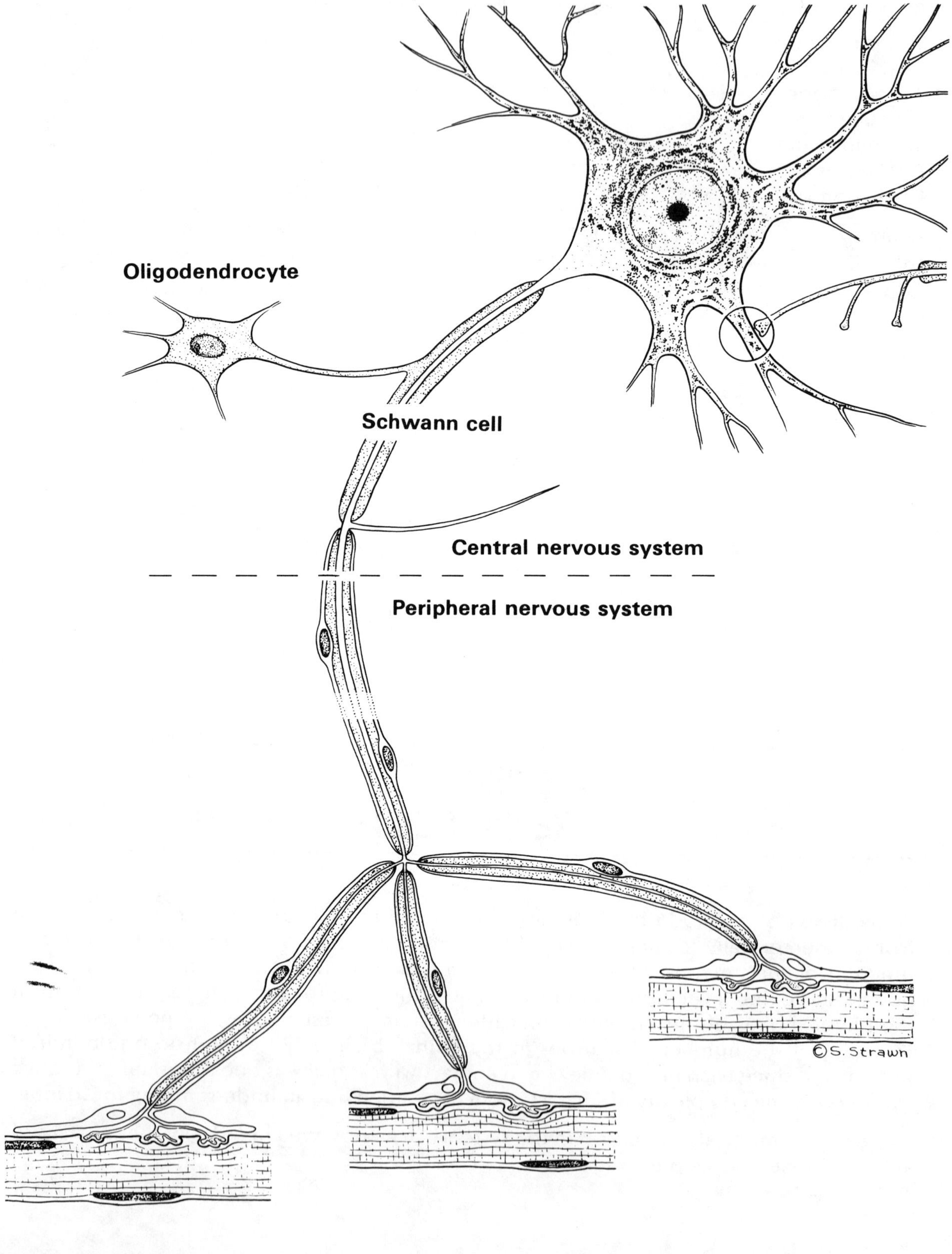

Figure 3-6. Four consecutive phases of myelin formation in peripheral nerve fibers.

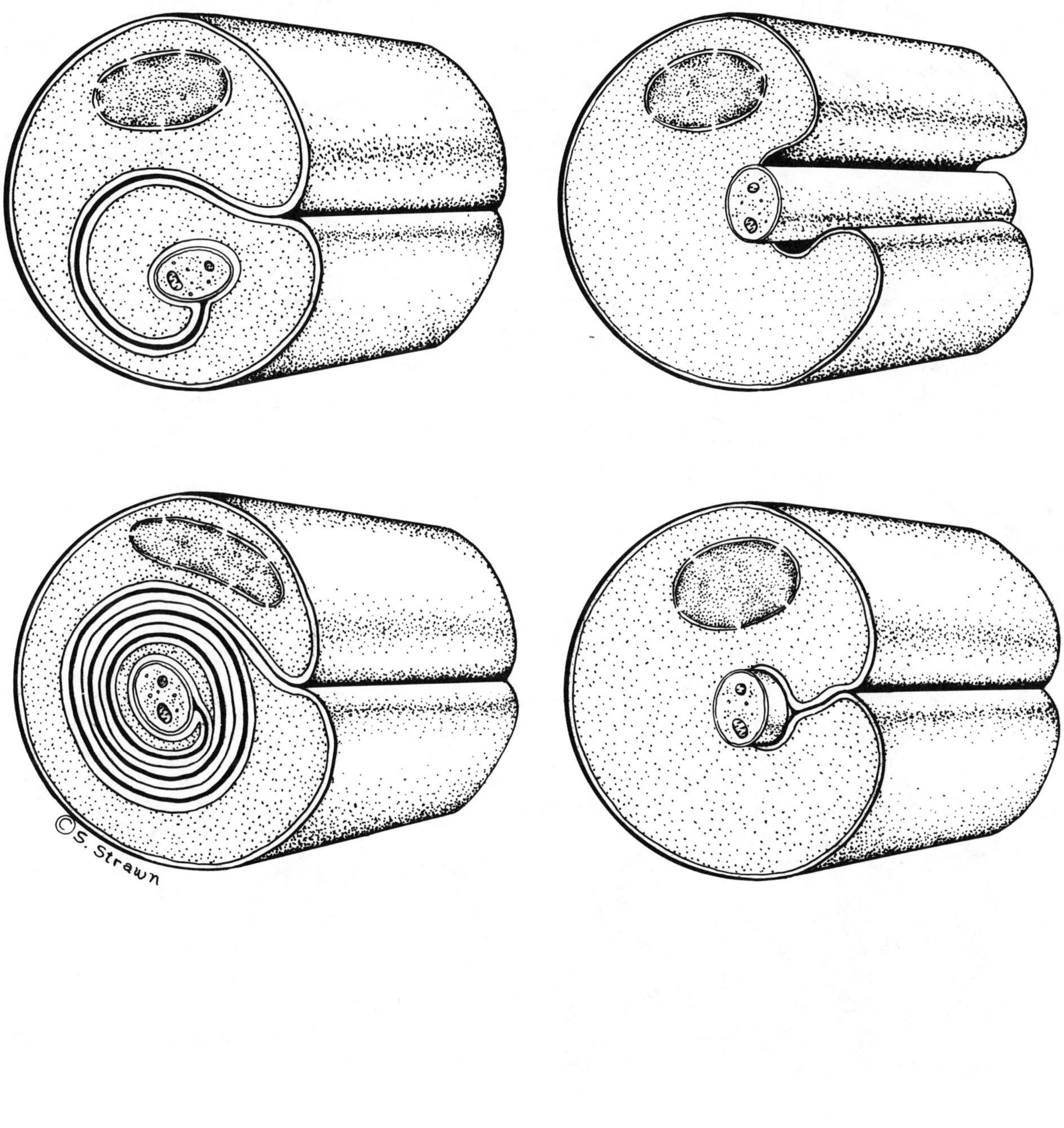

Figure 3-7. Two arrangements of unmyelinated fibers in a supporting Schwann cell.

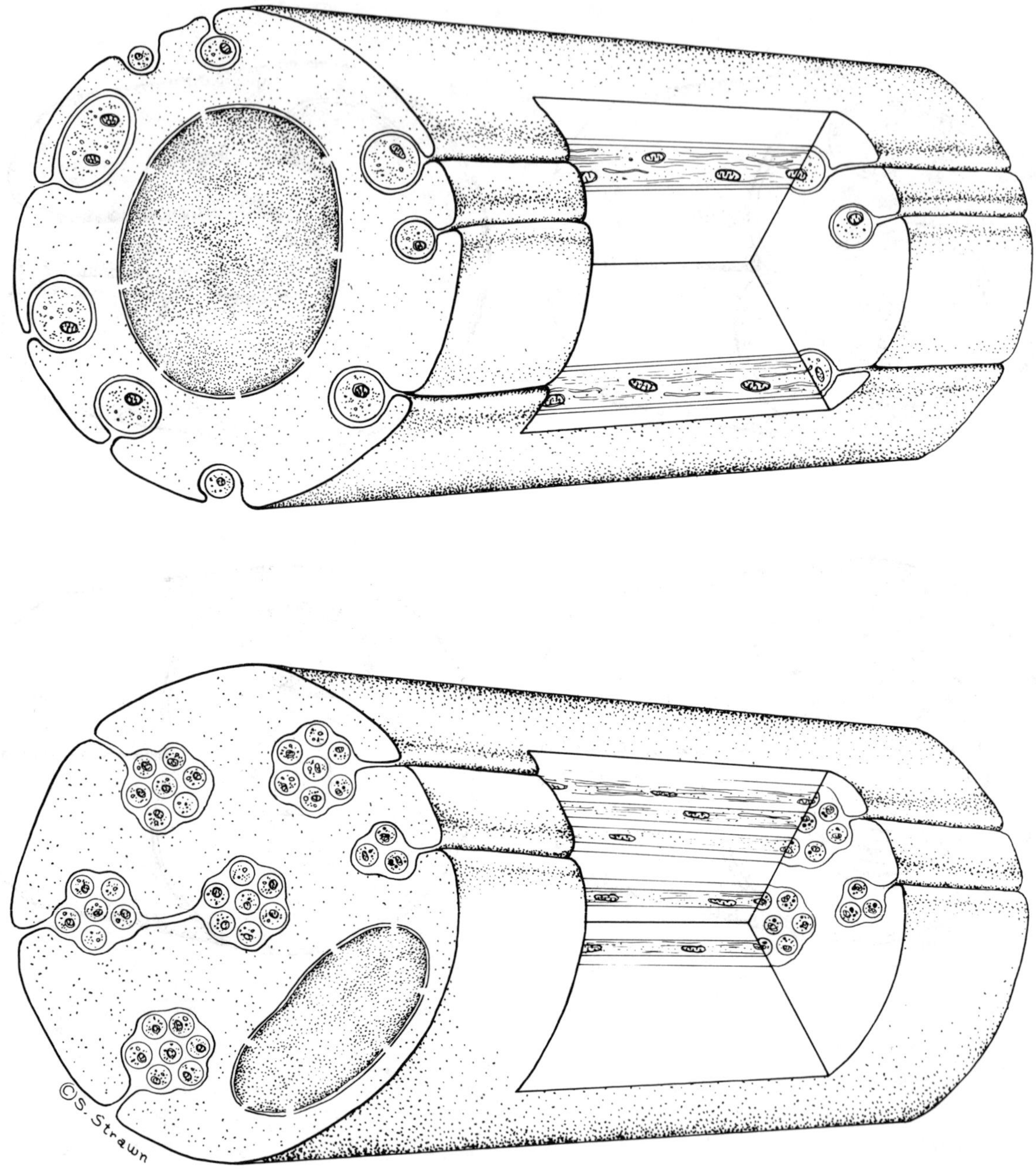

Cross Section of the Spinal Cord

Place a slide containing a cross section of the spinal cord on your microscope and locate as many of the following structures as possible. Label these structures on Figure 3-5.

1. Cell body
2. Dendrites
3. Axon
4. Nucleus of cell body
5. Axon
6. Motor end plate
7. Muscle fiber
8. Synapse

Teased Nerve Fibers

Place a slide containing a teased nerve fiber on your microscope and locate and sketch the following structures.

1. Myelin sheath
2. Nodes of Ranvier
3. Internode
4. Schwann cell

Figure 3-6 shows the process of myelination of a myelinated nerve fiber. The large Schwann cell forms a groove into which the axon moves. Through a process of successive wrappings around the axon, much of the cytoplasm of the Schwann cell is squeezed out, and many layers of cell membrane surround the axon at the internode. Each Schwann cell in peripheral nerves forms one internode along the axon.

Figure 3-7 illustrates two arrangements on unmyelinated fibers in a supporting Schwann cell. In the top view, each axon is enveloped in its own "tunnel;" in the lower view, several unmyelinated fibers occupy the same "tunnel."

What is the functional significance of myelination in the myelinated nerve fibers?

____

____

LABORATORY 4

The Skeletal System (General)
The Axial Skeleton

OBJECTIVES

This laboratory period will enable you to learn the following skeletal structures.

1. Axial Skeleton

a. The Skull

Bones	*Features*
Frontal	Orbital Cavity
Temporal	Auditory Meatus
Mandible	Foramen Magnum
Zygomatic	Coronal Suture
Parietal	Zygomatic Arch
Maxilla	Mastoid Process
Nasal	Occipital Condyles
Occipital	

b. The Vertebral Column

Body
Pedicle
Vertebral Foramen
Superior Articular Process
Spinous Process
Inferior Articular Process

c. The Rib Cage

Ribs
Sternum

2. Appendicular Skeleton

a. Upper Limb

Clavicle
Scapula
Humerus
Radius
Ulna
Carpals (as a group)
Metacarpals (as a group)
Phalanges (as a group)

b. Lower Limb

Ischium
Ilium
Pubis
Femur
Fibula
Tibia

The skeleton consists of the bones of the body and the joints and ligaments that hold the bones together. The skeleton is the framework upon which the softer tissues depend for support and protection. Its bones form levers that greatly improve the capability of muscles to accomplish a wide variety of forceful, rapid, and precise movements. The shape and size of the bones are highly specialized for the kind of movements required by any species in its attempt to survive. The bones also provide a storage site for calcium and phosphorus, which are necessary for many functions of all tissues. Finally, the red blood corpuscles and certain of the white blood cells are formed in the red bone marrow inside some of the bones.

Although some form of skeleton is found in many of the lower animal phyla, it is only in the arthropods and vertebrates that we have a true, jointed skeleton that significantly improves the animals' capacity for movement. A look at the number of species and the vast range of niches occupied by these species testifies to the importance of the skeleton in the success of a phylum.

Knowing the parts of the skeleton is the first step toward knowing the other anatomical structures of the body, since many of them are named for the skeletal parts that support them or which they move. For example, most arteries, veins, and nerves are named for the bones nearest them. Thus, we have the radial artery and vein, the tibial nerve, the ulnar artery and vein, and many others. Movements of the body are described in terms of joint actions, such as flexion of the elbow, extension of the hip, rotation of the head, and others. An understanding of all of gross anatomy and the movements of the body starts with knowledge of skeletal structures.

Along with the muscular system, the skeletal system is highly specialized for the varieties and range of movement of different parts of the limbs, trunk, and head. The actions that a cat can perform compared to those that a human can perform are entirely dependent on the shape, position, and relative size of the bones and the muscles that are attached to those bones. The major differences that we will see between the human skeleton and the cat skeleton are primarily the result of the evolution of bipedal locomotion by man and the retention of four-footed or tetrapodal locomotion by the cat. For example, the relatively larger facial bones of the cat and the muscles attached thereto are responsible for a number of actions for which the upper limb of man has been adapted. Fighting, catching prey, and carrying objects are primary functions of the facial bones and muscles of the cat and of the upper limbs of man and his ancestors. The vertebral column of the cat forms an arch that not only functions to support the weight of the internal viscera, but that is capable of extreme flexion and extension to increase the velocity of running. In man, the vertebral column has four distinct curvatures that enable it to act as a spring to absorb shock in various forms of locomotion, such as walking, running, and jumping. Furthermore, the size of the vertebrae increases from top to bottom due to the requirement of weight bearing in bipedal locomotion. The overall shape of the cross section of the rib cage in the cat is roughly heart shaped, whereas that in man is much wider than it is deep. This causes a difference in the location of the scapulae (the shoulder blades), which accounts for major differences in the use of the upper limb, as we will see when we come to that laboratory period. Details of the limb skeleton will be discussed in subsequent laboratory sessions.

EXERCISE 1
ASSEMBLY OF THE DISARTICULATED HUMAN SKELETON

From the bones supplied on your laboratory tables, assemble an entire skeleton, placing each of the bones in its appropriate location. Use your text and the articulated skeletons as aids in this assignment.

How are you able to distinguish between right and left bones?

____

____

____

____

EXERCISE 2

Label the following bones on Figure 4-1. Use your texts for help.

Axial Skeleton

1. Skull—frontal bone, parietal bone, temporal bone, maxilla, mandible
2. Cervical vertebrae, thoracic vertebrae, lumbar vertebrae, and sacrum
3. Rib Cage—first rib, twelfth rib, sternum

Appendicular Skeleton

1. Upper Limb—scapula, clavicle, humerus, radius, ulna, carpals, metacarpals, phalanges
2. Lower Limb—ischium, ilium, pubis, femur, patella, tibia, fibula, tarsals, metatarsals, phalanges

Figure 4-1.

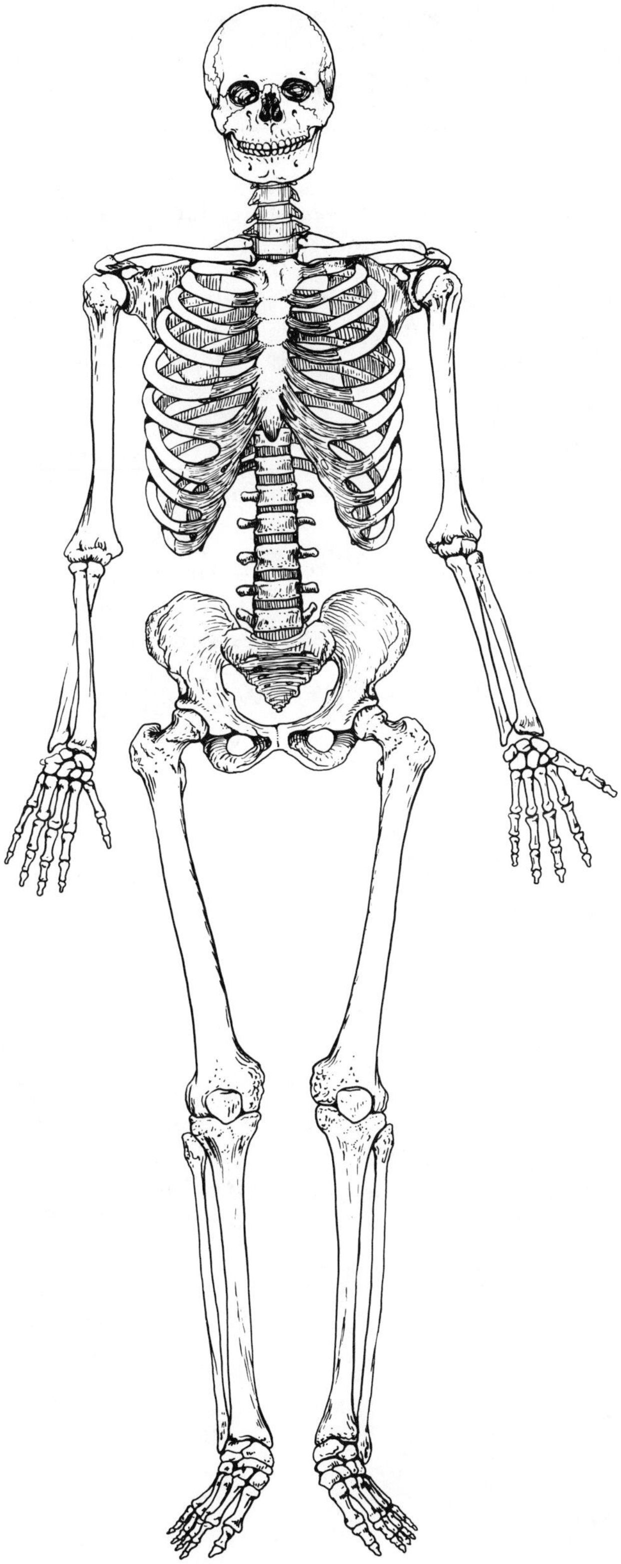

EXERCISE 3
THE SKULL

Using your text and the skeletal materials available, label the photographs on the next pages and answer the questions as indicated.

Examine the *front view* of the cat and human skulls on Figures 4-2 and 4-3. Label the following structures using the code below.

Bones

Man

F—Frontal
T—Temporal
Mn—Mandible
Z—Zygomatic
P—Parietal
Mx—Maxilla
N—Nasal
E—Ethmoid
S—Sphenoid
L—Lacrimal
V—Vomer

Cat

F—Frontal
T—Temporal
Mn—Mandible
Z—Zygomatic
P—Parietal
Mx—Maxilla
PMx—Pre-Maxilla
N—Nasal

Bone Markings

Man

Oc—Orbital cavity
Cs—Coronal suture
Za—Zygomatic arch
Mc—Medial conchae
Ic—Inferior conchae
MF—Mental Foramen

Cat

Oc—Orbital cavity
Ss—Sagittal suture
Za—Zygomatic arch
Mc—Medial conchae
MF—Mental Foramen

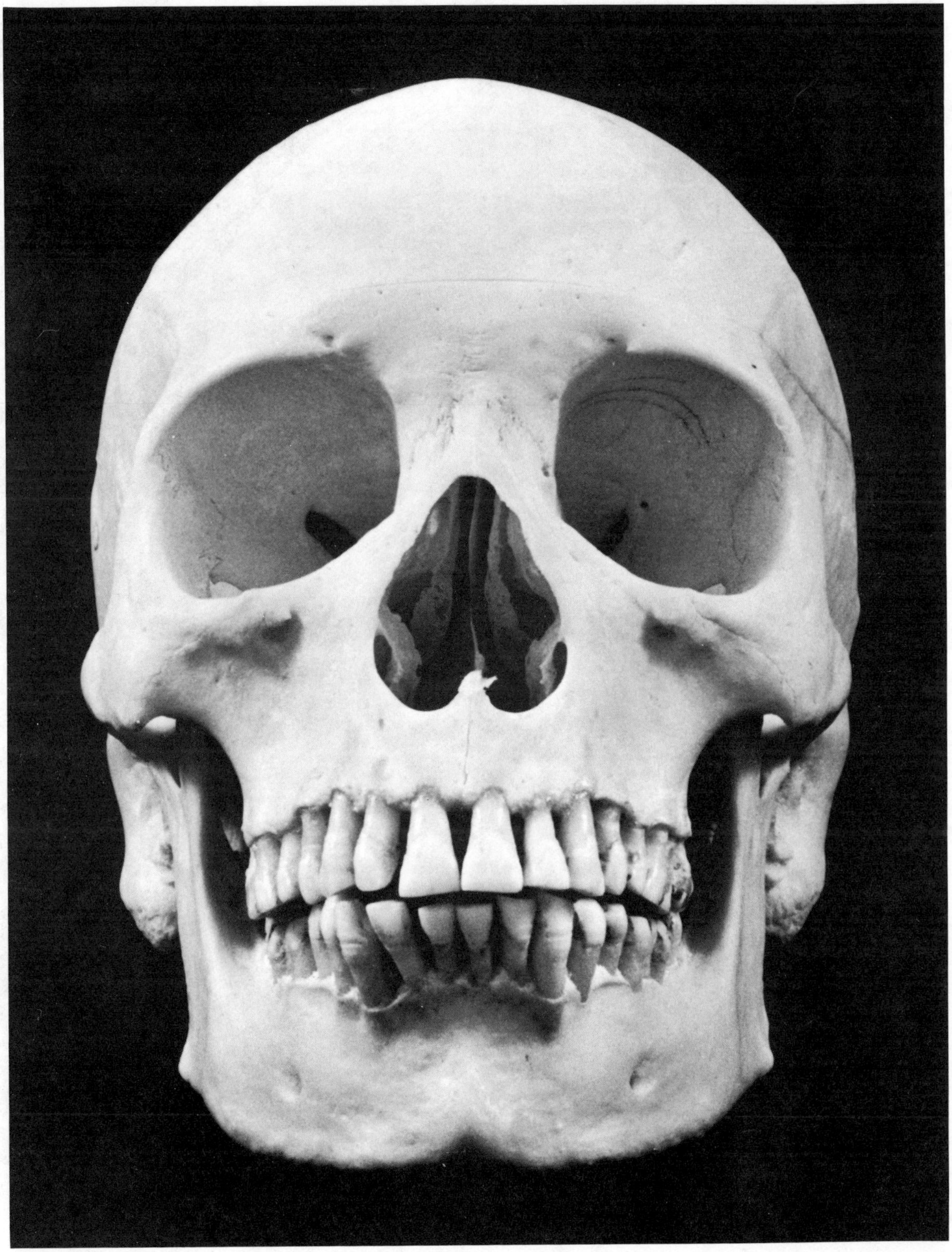

Figure 4-2.

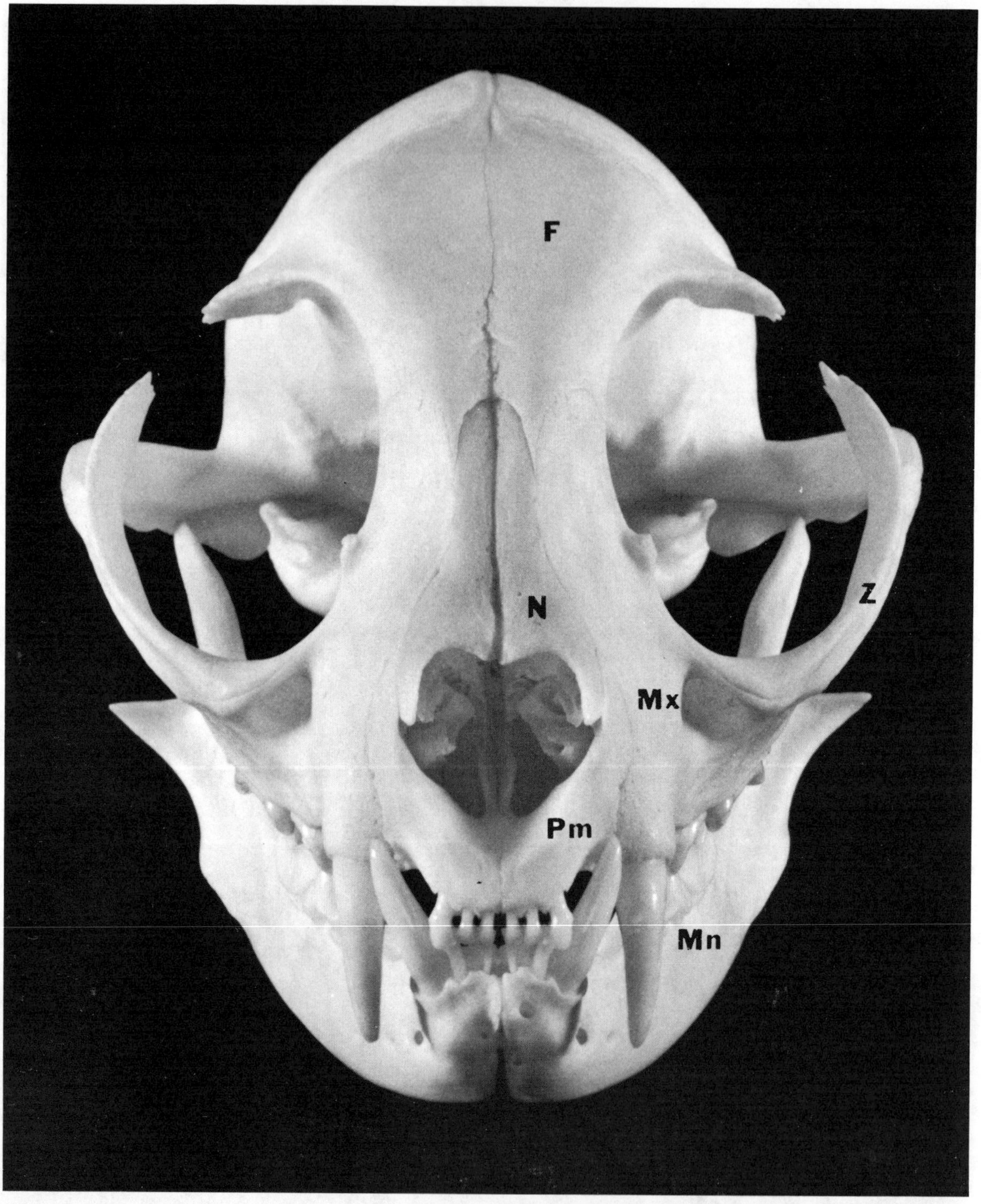

Figure 4-3.

Examine the *side view* of the cat and human skulls on Figures 4-4 and 4-5. Label the following structures using the code below.

Bones

Man

F—Frontal
P—Parietal
T—Temporal
O—Occipital
S—Sphenoid
E—Ethmoid
L—Lacrimal
N—Nasal
Z—Zygomatic
Mx—Maxilla
Mn—Mandible

Cat

F—Frontal
P—Parietal
T—Temporal
O—Occipital
S—Sphenoid
N—Nasal
Z—Zygomatic
Mx—Maxilla
Mn—Mandible

Bone Markings

Man

Za—Zygomatic process
Ea—External auditory meatus
MF—Mental Foramen
MP—Mastoid Process
Cs—Coronal suture
Ls—Lambdoidal suture
Sq—Squamous suture

Cat

Za—Zygomatic process
Ea—External auditory meatus
MF—Mental Foramen

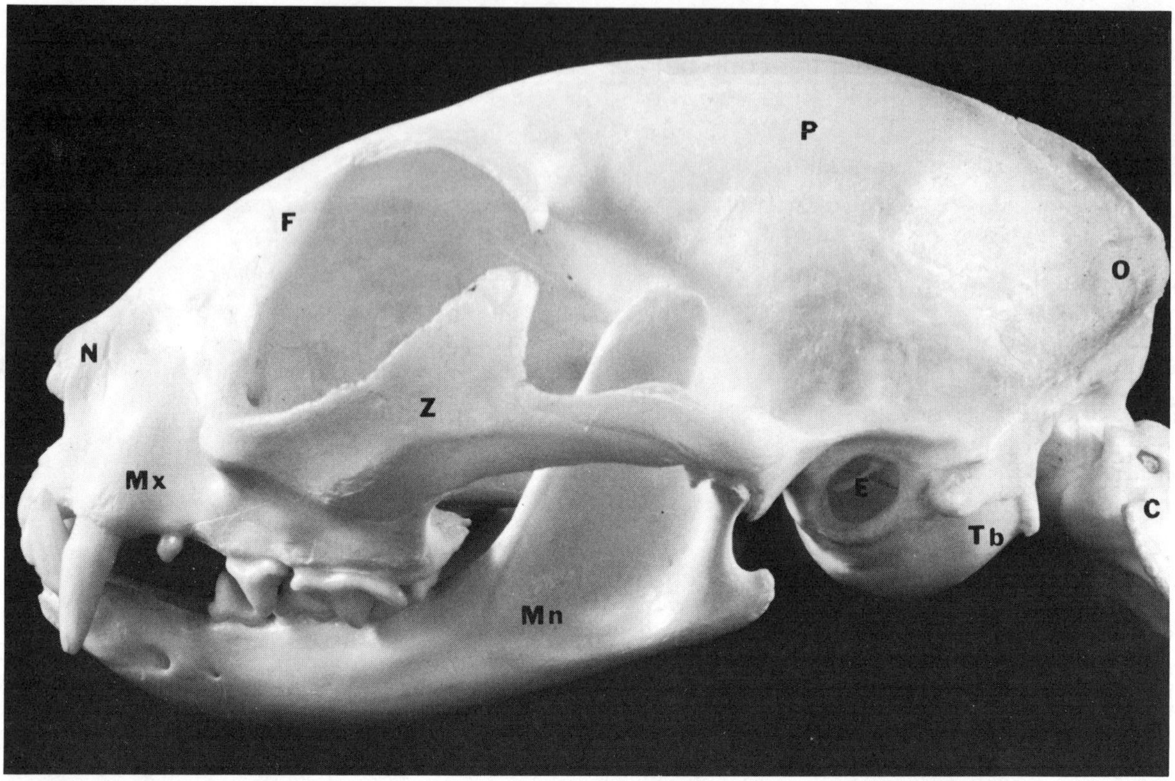

Figure 4-5.

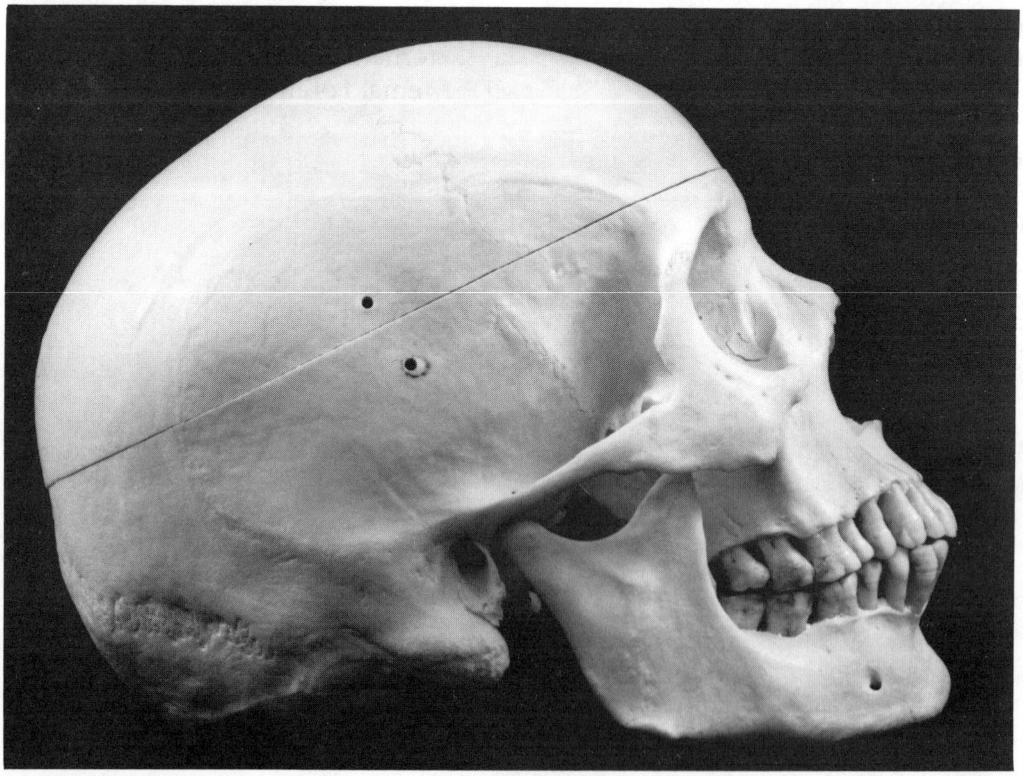

Figure 4-4.

Examine the inferior view of the cat and human skulls, mandibles removed, on Figures 4-6 and 4-7. Label the following structures using the code below.

Bones

Man

Mxpp—Maxilla palatine process
Pa—Palatine
Z—Zygomatic
S—Sphenoid
V—Vomer
O—Occipital

Cat

Mxpp—Maxilla palatine process
Pa—Palatine
Z—Zygomatic
S—Sphenoid
V—Vomer
O—Occipital

Bone Markings

Man

Occ—Occipital condyle
Fm—Foramen magnum
MP—Mastoid Process

Cat

Occ—Occipital condyle
Fm—Foramen magnum
Tb—Tympanic bulla

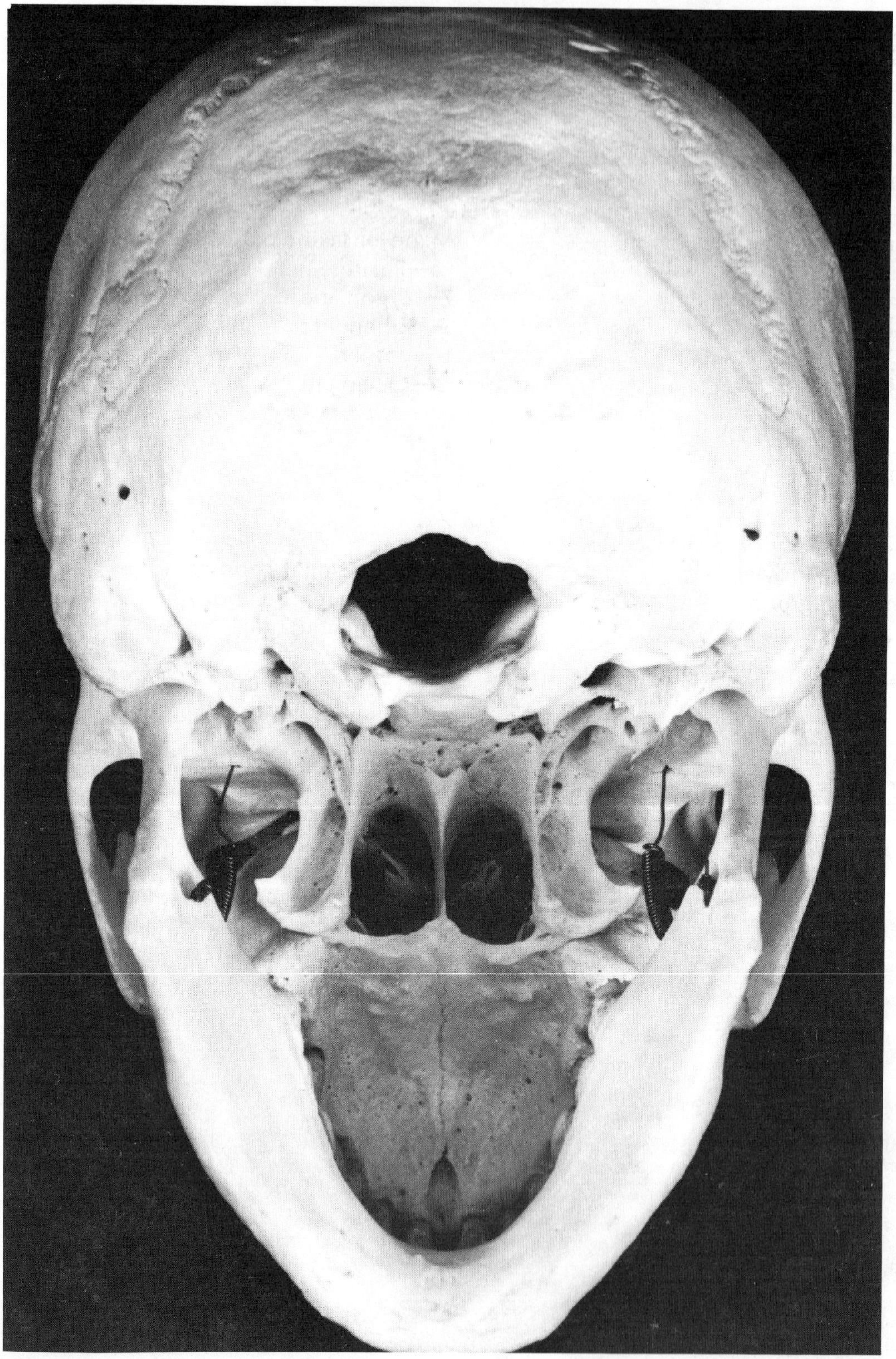

Figure 4-6.

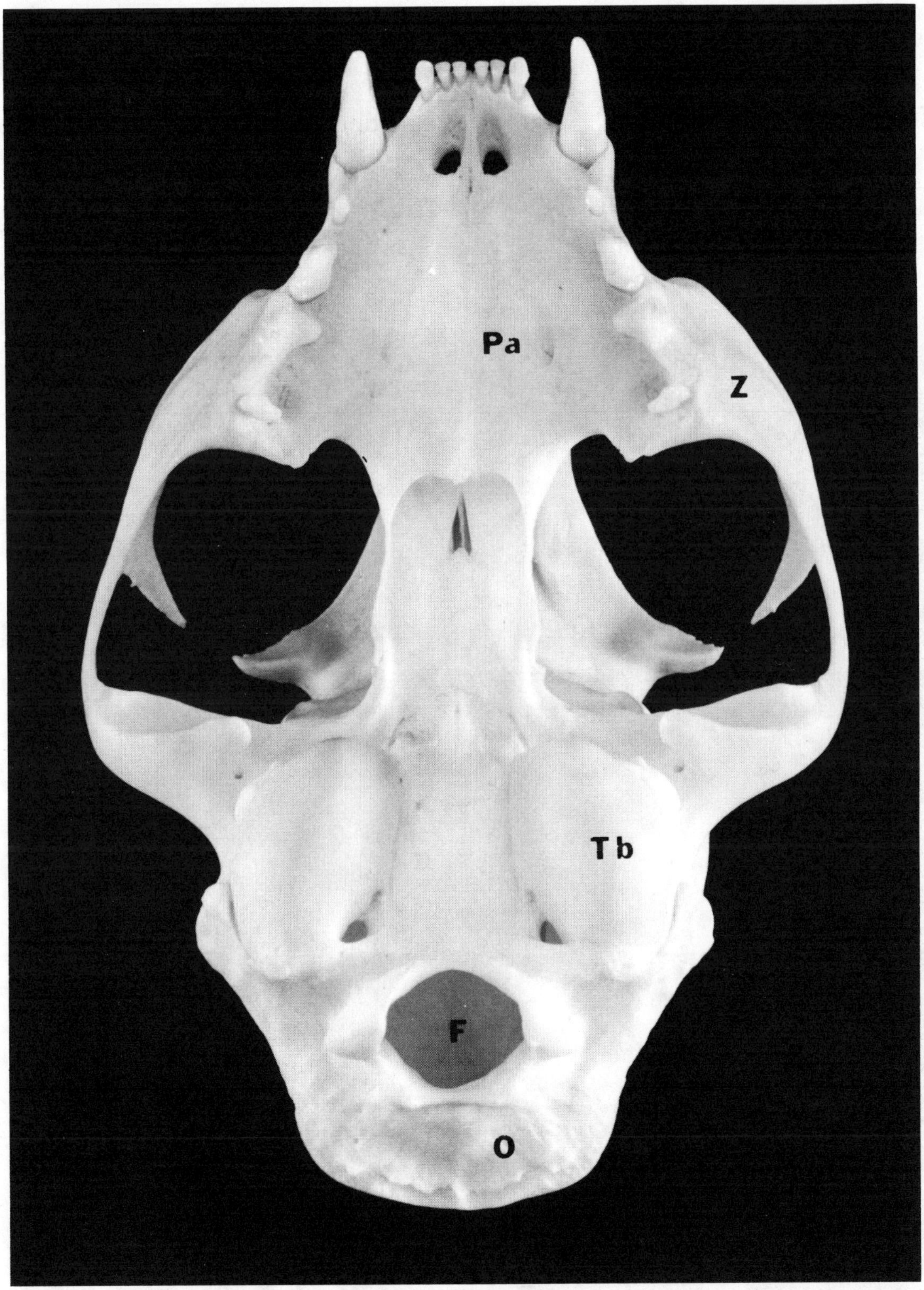

Figure 4-7.

Examine the top view of the human skull and label the following structures on Figure 4-8, using the following code.

Bones

F—Frontal
P—Parietal
O—Occipital

Bone Markings

Cs—Coronal suture
Ss—Sagittal suture
Ls—Lambdoidal suture

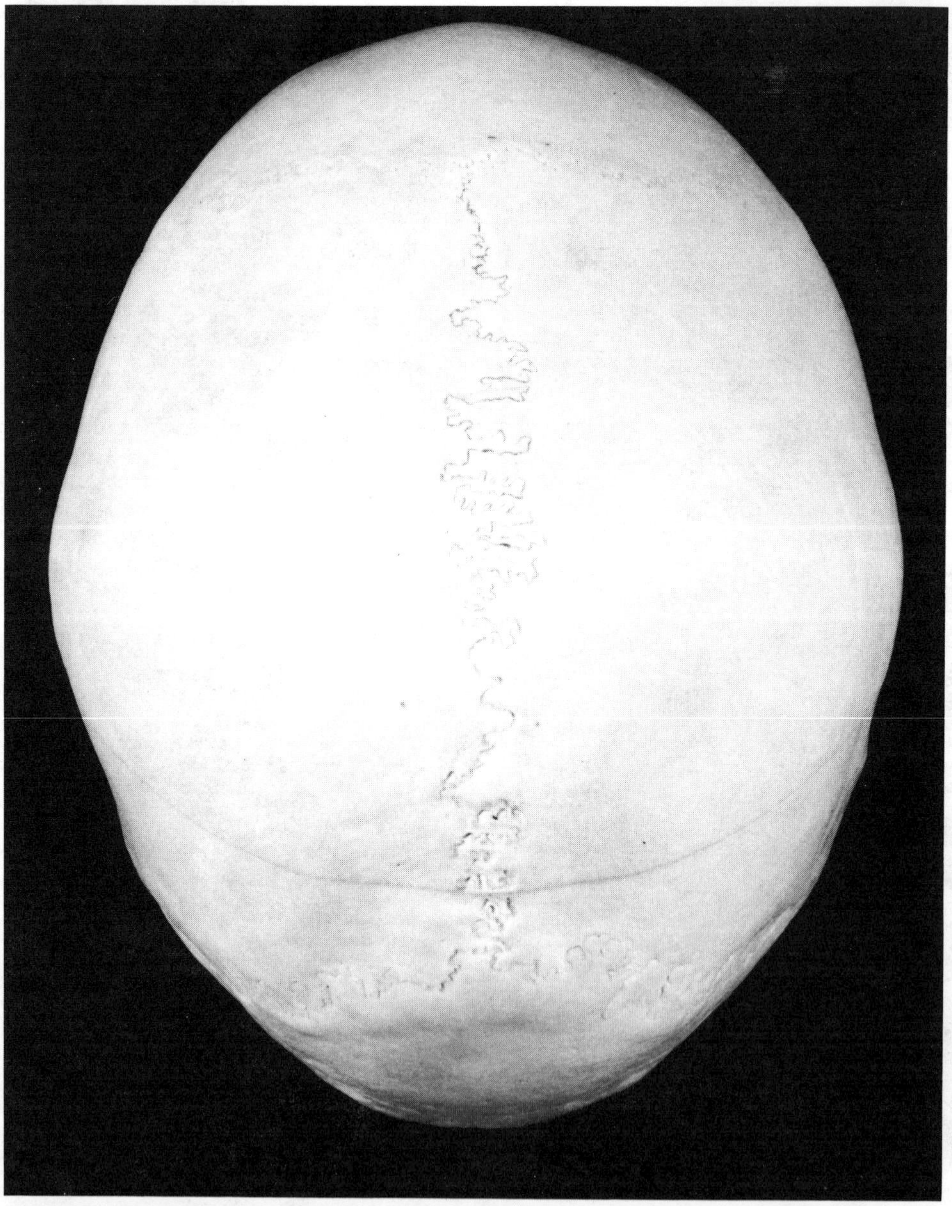

Figure 4-8.

EXERCISE 4
COMPARISON OF CAT AND HUMAN SKULLS

1. How does the skull of the human differ from that of the cat?

2. How can you account for the differences above?

3. What are some examples of organs protected by the skull?

4. Give examples of the support function of the skull.

5. For what muscular activities does the skull furnish bony leverage systems?

____

____

____

____

6. Locate the following structures on yourself by palpation.

Zygomatic arch
Angle of mandible
Mastoid process
Inion
Supra-orbital margin
Nasal bone

EXERCISE 5
THE VERTEBRAL COLUMN

Examine the vertebrae on the laboratory table. Locate the following structures and label them on Figures 0-0, 0-0, and 0-0. Use the following code.

B—Body
P—Pedicle
F—Vertebral foramen
Sa—Superior articular process
Sp—Spinous process
Ia—Inferior articular process

Complete Table 4-1 by filling in the required information.

Table 4-1

Region of Vertebral Column	Number of Vertebrae	Relative Thickness of Discs	Relative Mobility	Special Features
Cervical				
Thoracic				
Lumbar				
Sacral				

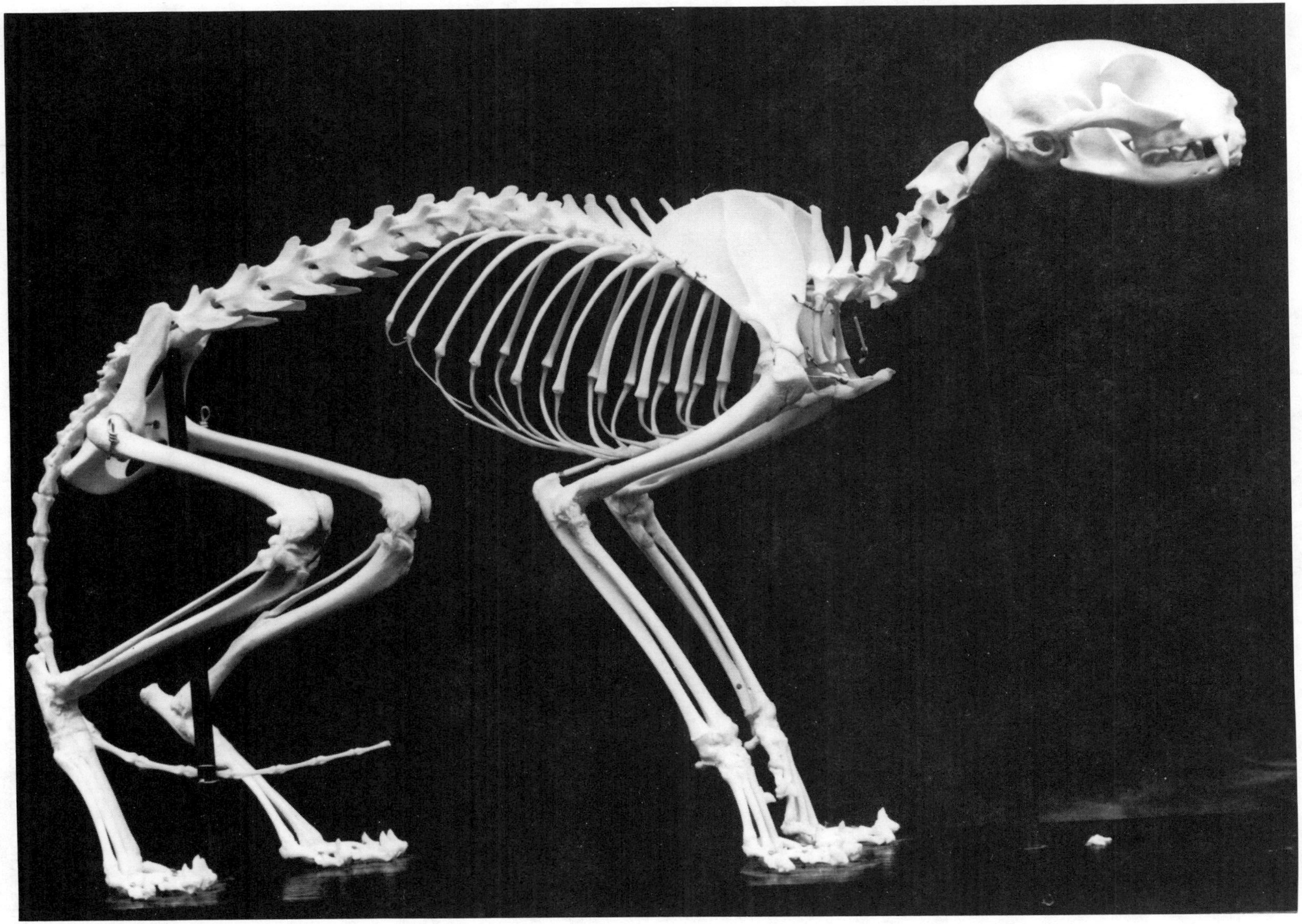

Figure 4-9. Cat skeleton.

In the space below, diagram the vertebral column of the human, showing the curves by a single line drawing. Do the same for the cat after examining the mounted skeleton or Figure 4-9.

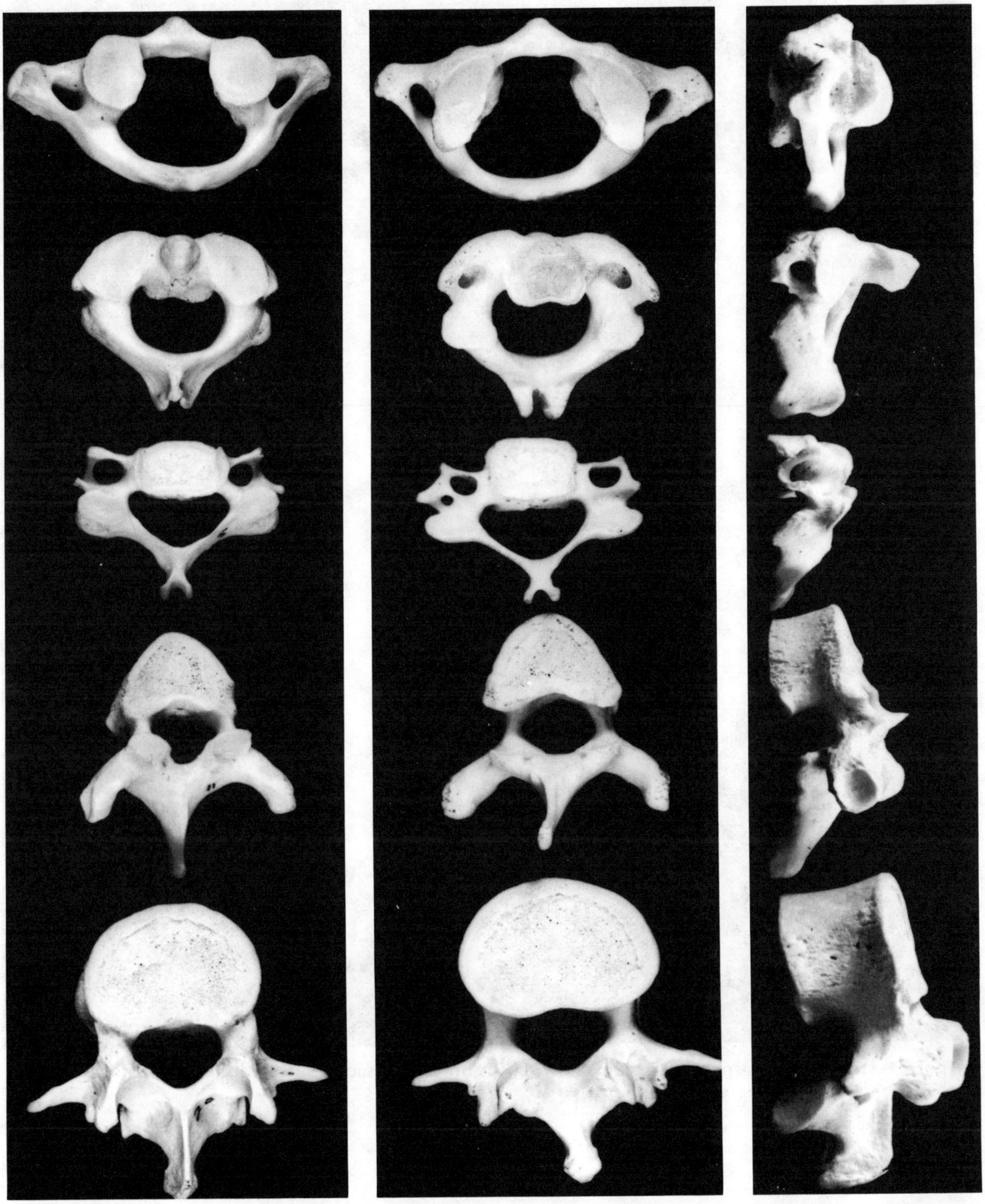

Figure 4-10. A—superior aspect; B—inferior aspect; C—lateral aspect. From top to bottom, first cervical vertebra (atlas), second cervical vertebra (axis), sixth cervical vertebra, fourth thoracic vertebra, second lumbar vertebra.

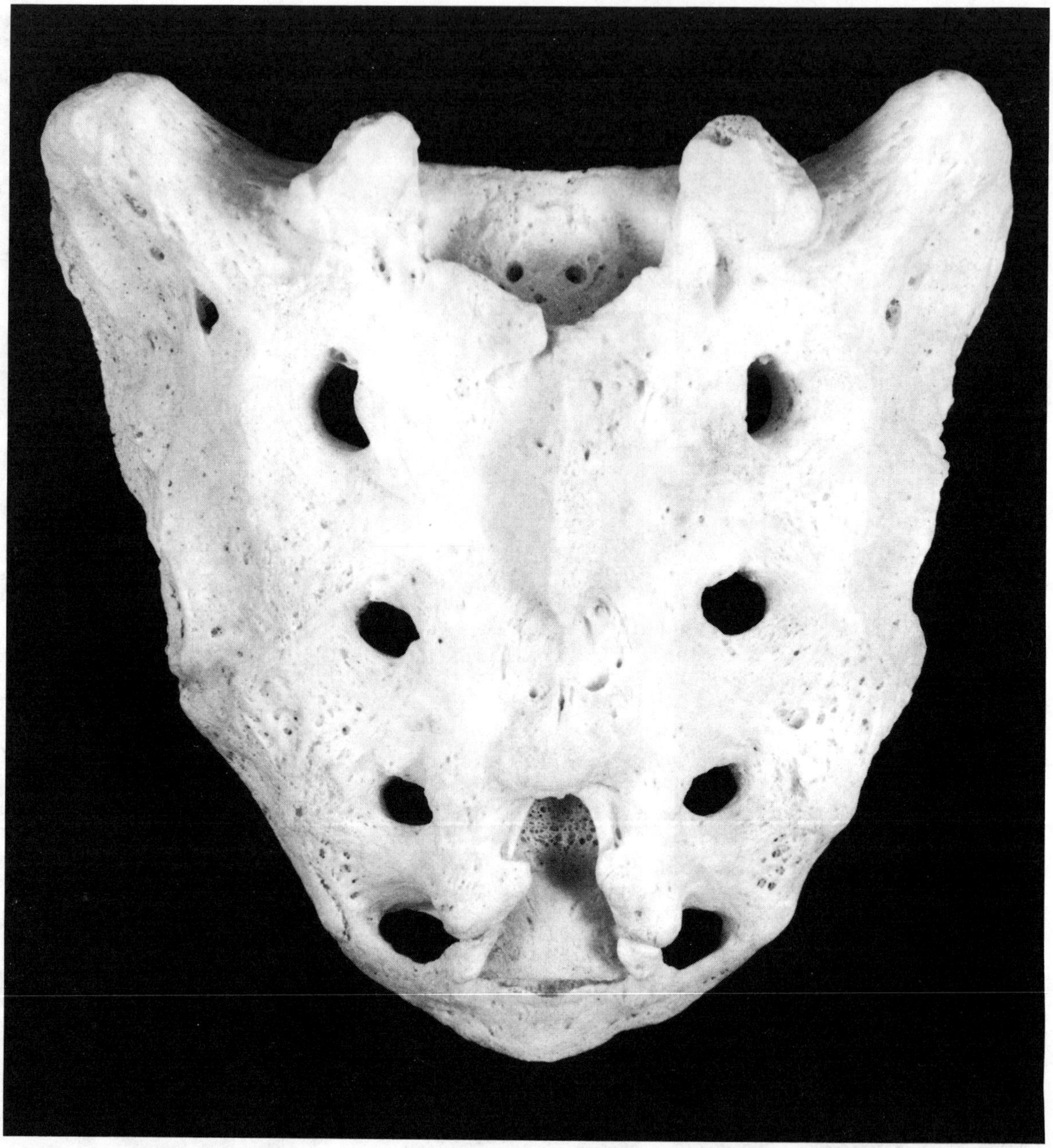

Figure 4-11. Posterior aspect of sacrum showing fusion of five sacral vertebrae leaving foramena from which spinal nerves emerge.

1. Contrast the function of the vertebral column in the human to that of the cat.

2. What protective function does the vertebral column fulfill?

3. What adaptations for effective leverage in maintaining an erect posture do you find in the human vertebral column?

EXERCISE 6
RIB CAGE

Count the number of ribs on a human skeleton and on a cat skeleton. How many are in each?

Human____________________ Human____________________

Cat____________________ Cat____________________

Measure the width in centimeters of the rib cage in both the cat and human skeletons at the widest part.

Human____________________

Cat____________________

Measure the anterior-posterior distance in both the human and cat skeletons at the same level as before.

Human____________________

Cat____________________

What is the ratio of width to depth in each?

Human____________________

Cat____________________

In the space below, sketch a general outline of a cross section of the middle of the human rib cage. Do the same for the cat. Draw a line representing the position of the scapulae on each.

How do the positions of the scapulae on the rib cage affect the use of the forelimbs in man and the cat?

____

____

____

LABORATORY 5

Anatomical Terminology
Dissection Techniques

OBJECTIVES

Following this laboratory, you should be able to accomplish the following objectives.

1. Understand anatomical terminology, including terms of direction, fundamental planes, and the anatomical position.
2. Understand and demonstrate the proper use of instruments and techniques in dissection.

EXERCISE 1
TERMS OF REFERENCE

With the aid of your textbook and instructor, define each of the following terms in a few words.

1. Superior ____________________

2. Inferior ____________________

3. Anterior ____________________

4. Posterior ____________________

5. Ventral ____________________

6. Dorsal ____________________

7. Caudal ____________________

8. Cranial ____________________

9. Medial ____________________

10. Lateral ____________________

11. Superficial ____________________

12. Deep ____________________

EXERCISE 2

It is clear that some of these terms of reference are synonymous. However, the difference in position between bipedal man and the tetrapodal cat causes some confusion in which of the terms are interchangeable. Place the following terms in the appropriate eight spaces in the drawing of the man and cat.

Superior
Inferior

Cranial
Caudal

Ventral
Dorsal

Anterior
Posterior

Figure 5-1. Directional terms for the cat and man.

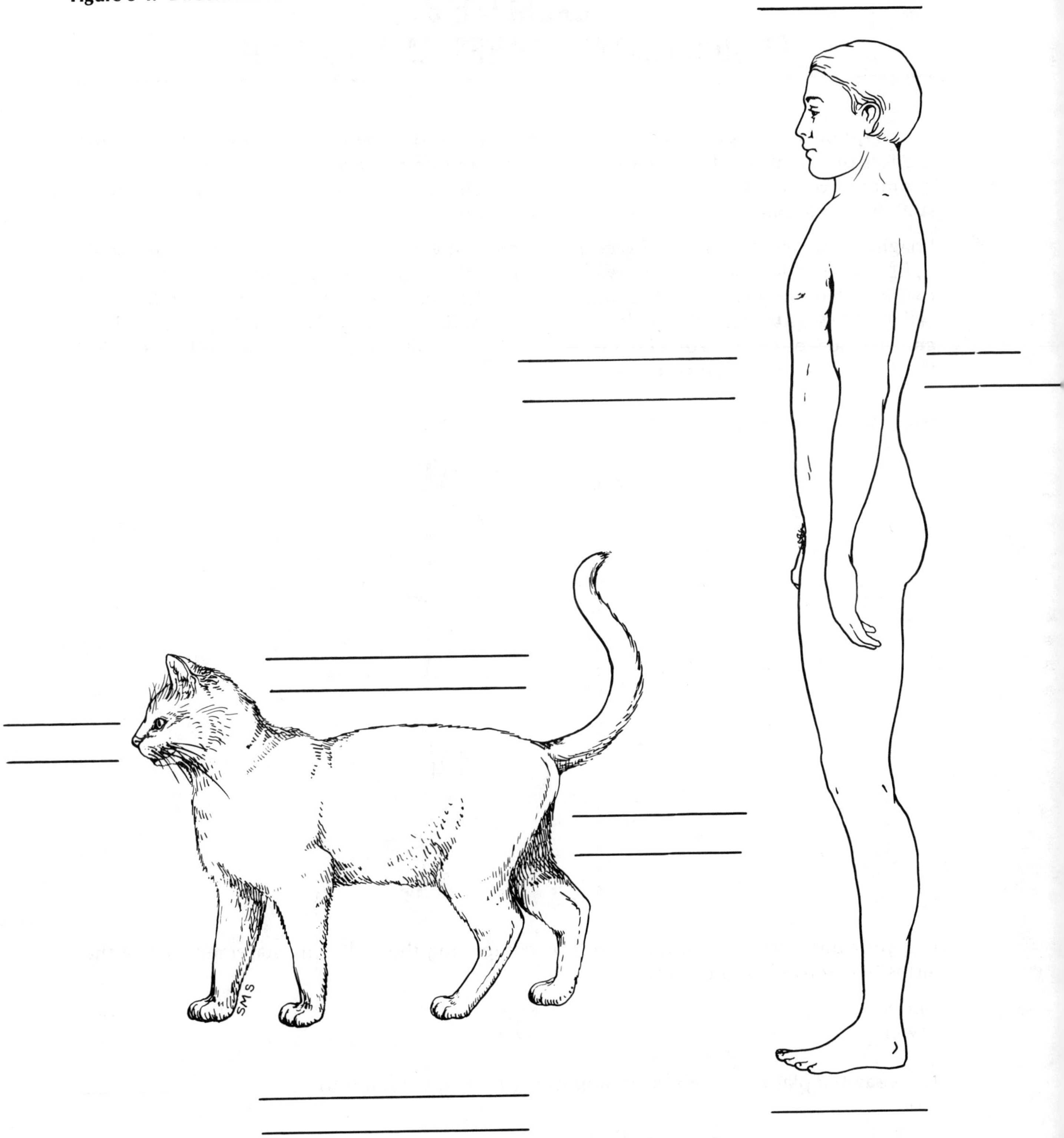

EXERCISE 3
FUNDAMENTAL PLANES AND SECTIONS

It is frequently necessary to refer to imaginary planes of reference in order to describe the motion of a limb or the surface of a cut portion of the body. If you remember your geometry, you know that a plane is a flat or uncurved surface so oriented in space that a straight line joining any two points on it lies entirely within it.

Imagine yourself standing in the corner of a room with your back to one wall, and your right side touching the adjoining wall. The three planes of your body are then each parallel to one wall or to the floor and ceiling. A plane that is parallel to the wall at your right side is called "the **sagittal** (saj"-it'-tal) plane;" a plane parallel to the wall at your back is called the **coronal** (cor'-o'-nal) or **frontal** plane; and a plane parallel to the floor and ceiling is called the **transverse** or **horizontal** plane.

Figure 5-2. Fundamental planes.

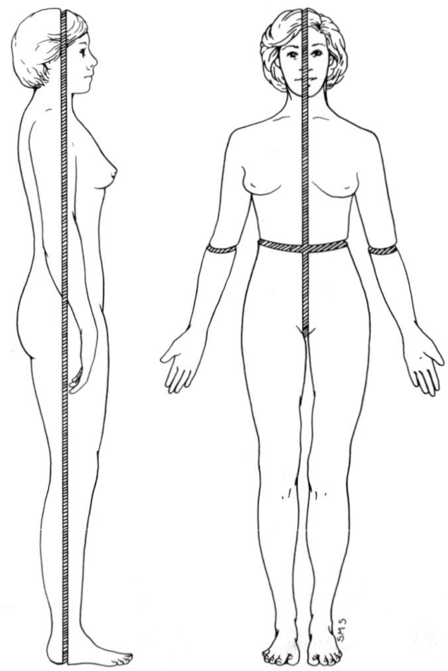

Test your understanding of the planes by completing the following statements using the terms below to fill in the blanks.

Upper	Right	Front
Lower	Left	Back

1. A **sagittal** plane through the center of the body divides it into __________________

   and __________________ parts.

2. A **coronal** or **frontal** plane through the center of the body divides it into ____________ ____________ and ____________ halves.

3. A **transverse** or **horizontal** plane through the center of the body divides it into ____________ ____________ and ____________ parts.

EXERCISE 4
THE ANATOMICAL POSITION

The figure in the drawing is shown in the anatomical position. In the space below, describe this position with special reference to the head, limbs, and hands.

Figure 5-3. The anatomical position.

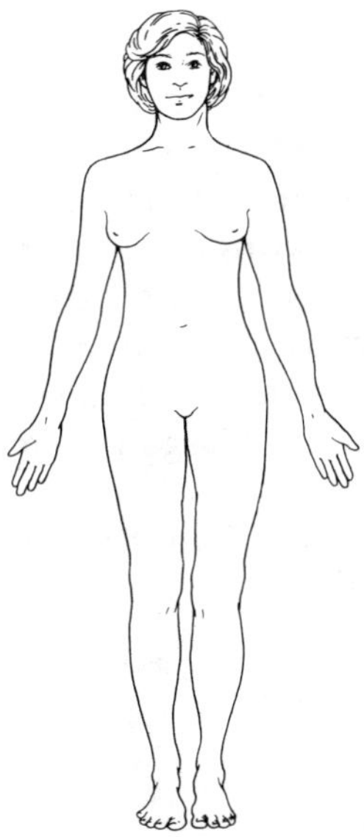

Head ____

____

Limbs ____

____

Hands ____

____

DISSECTION OF CAT MUSCULATURE

The following laboratory periods are devoted to dissecting most of the musculature of the cat. The laboratory assignments are purposely left flexible so that differences in dissecting ability and variations in ease of dissecting individual cats can be accounted for.

It is more profitable to dissect nerves, blood vessels, and muscles at the same time region by region. Therefore, although the emphasis in these following laboratory periods is on muscles, it is necessary to know the skeletal structures, nerves, and blood vessels at the same time.

General Instructions for Dissection

Dissection is the process of cutting apart and separating the various parts of an animal body so that they can be observed in detail and their relationships can be analyzed. The purpose of dissection is to attempt to understand as much as possible of the function of structures by thorough study of their form.

The major task in dissection is the process of "clearing" or removing the connective tissue wrappings from the organs being examined. In order to study organs lying deep within the body, it is frequently necessary to "transect" more superficial structures. This is done by freeing the muscle, ligament, or tendon from the underlying organs and cutting it across the center, leaving the two ends attached to their original positions. It is important to transect only those structures that interfere with the investigation of underlying materials; therefore, cut only the minimum indicated in the detailed instructions for the dissection of a given region. Never remove an organ or organ system until it has been completely studied and until you are sure that you will not have to refer to it again. Removal of any structure should be done only on direction from the laboratory instructor. Neatness and thoroughness are imperative. The quality of a dissection can be partially judged by its appearance.

Removal of Skin

1. Incisions in the skin of the cat are most easily, rapidly, and safely made with a pair of heavy scissors, with one blunt blade. Cutting with a scalpel or sharply pointed scissors is often the cause of destruction of important structures lying just under the skin. In making an incision, pinch a small section of skin between the thumb and forefinger, and make a short cut at right angles to the desired line of the incision. Then, with the blunt blade down, cut along the incision line the required distance. Usually it will be necessary to make lateral cuts at an angle to the main incision in order to uncover a particular region.
2. After making the incision, lift the edge of the skin and scrape the underlying **fascia** away with a blunt instrument, *not the scalpel.* Uncover only the area that is to be studied. Leave the skin attached to the cat so that it can be used to cover the dissected

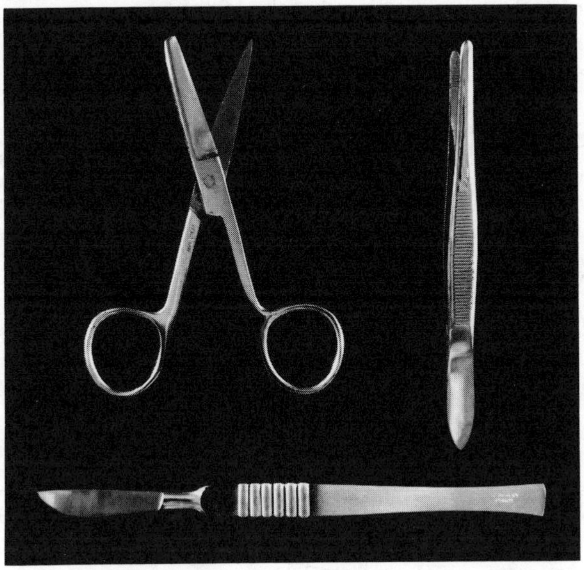

Figure 5-4. Dissecting instruments. A scalpel, strong forceps, and scissors with one blunt end are the only dissecting instruments, outside of your fingers, that you need for dissecting the cat as described in this manual. Special instruments such as bone saws will be provided by the instructor.

part and keep it from drying out. An extensive set of thin muscle fibers is attached to the skin on its interior surface and is removed with the skin. This muscle is the **cutaneous maximus** or the **panniculus carnosus.** It originates near the humeral attachment of the **latissimus dorsi**, and the fibers of the two are intricately interwoven in that region. Therefore, care must be used not to destroy portions of the **latissimus dorsi** in the skinning process.

Dissection of Muscles

1. When the skin has been removed, carefully scrape the surface of the exposed muscles with the scalpel blade. Clear the surface of fat and fascia using scalpel, forceps, and fingers. As the clearing progresses, the boundaries of the muscle will appear, and the general shape and location will be observable. A muscle should be separated completely from underlying and adjacent organs by inserting the blunt end of the scalpel or the fingers under and around it until the connective tissue that attaches it to surrounding structures is torn and the muscle is isolated. Trace the muscle to its skeletal attachments, and observe the direction and arrangement of the fibers — whether they are parallel, pennate, etc. Wherever possible, examine the mechanical arrangement of the muscle, its tendons, and the bones to which it is attached. Look for leverage types, relative length of lever arms, and similar cues to the importance and the function of the muscle. Much can be inferred from the bulk of the muscle, its fiber arrangement, and its mechanical linkage to the skeleton.

Dissection of Nerves

1. The peripheral nerves (cranial and spinal) usually lie in the connective tissues surrounding a muscle. Often they are accompanied by blood vessels. The main branch of a nerve gives off small branches that penetrate the muscle and, in turn, give off

Figure 5-5. Suggested sequence for incisions in skinning the cat, assuming that the lower limb is to be studied first.

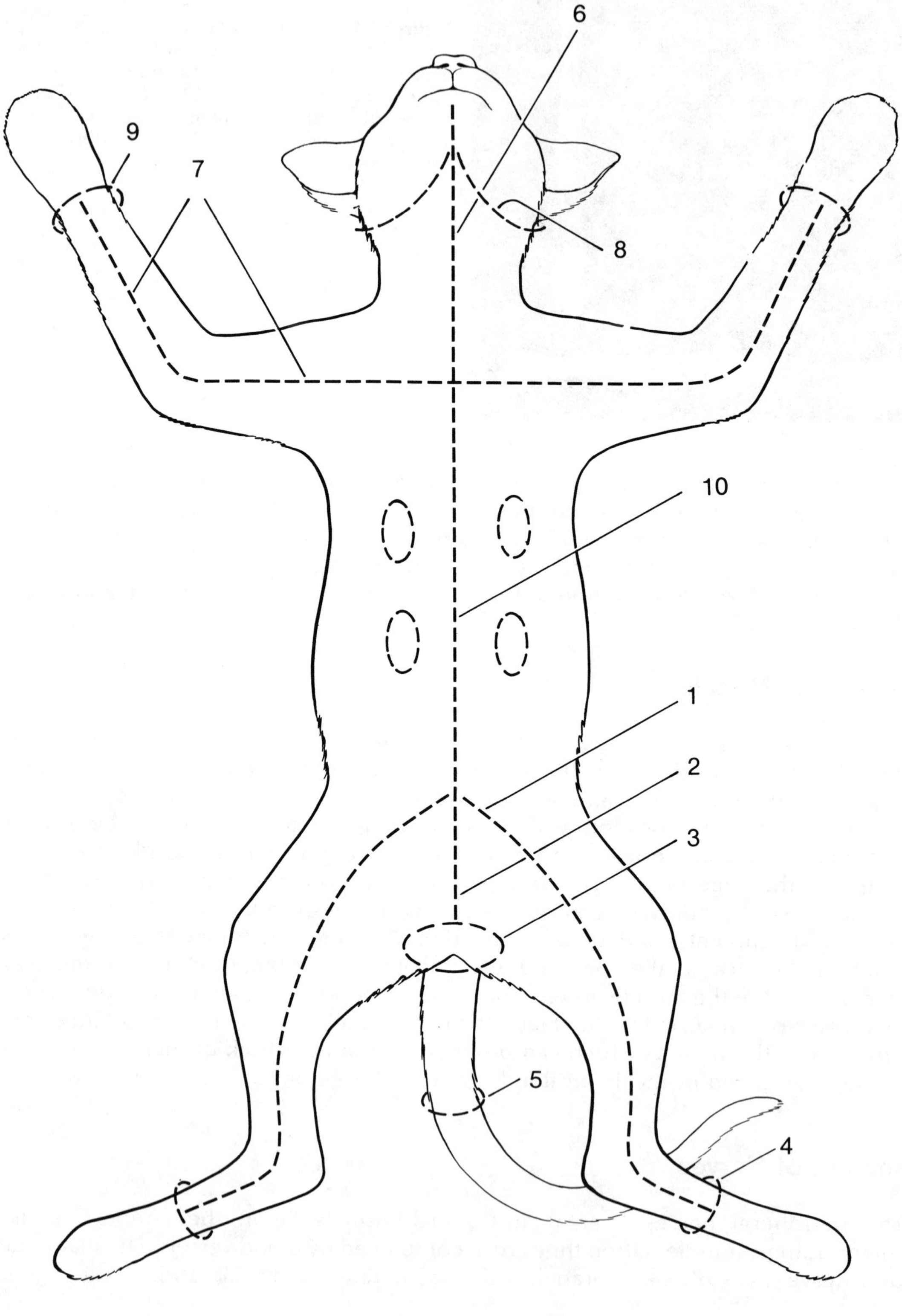

smaller branches to the motor units. These latter are microscopic in size and cannot be observed in a gross dissection. The connective tissue sheath (epinerium) must be removed before the location of branches can be closely observed. This is accomplished by splitting the sheath and running a blunt pair of forceps, or a probe, along the length of the nerve, using care not to cut through or rip off the smaller branches as they emerge from the main nerve trunk. Once the nerve is dissected, it should be traced to its origin, and its distribution to skin or muscle should be observed.

Dissection of Blood Vessels

1. The arteries of all cats used in this course will be injected with red latex. One or two cats in each laboratory will have the veins injected as well. Veins will be blue in color. Capillaries, of course, are too small to be observed with the unaided eye. Clearing of blood vessels requires only the removal of the thin sheath of connective tissue that surrounds the vessel. Frequently, an artery, a vein, and a nerve will be invested by the same connective tissue sheath and can be dissected merely by slitting the sheath and tracing the arteries to their origin from the **aorta**, and the veins to their junction with the **vena cava**. Muscular branches of the arteries and veins can be seen as branches from the main trunk. Variations are common in blood vessels. The veins in particular frequently deviate from the usual pattern.

LABORATORY 6

Dissection of Cat Lower Limb I: Hip and Thigh

In man, the hind limb is a highly specialized structure adapted for bipedal locomotion — one of the most significant characteristics of man. The limb shows many adaptations for stability and for the maintenance of erect posture. The cat, on the other hand, is a relatively nonspecialized or representative type of mammal, viewed in light of structural specializations. Phylogenetically, it is close enough to man to show many similarities of structure, which will be observed throughout the course of the dissection. The cat differs from man in one important aspect: it is a **tetrapod**, that is, it is a four-legged animal that walks and runs on all four limbs. More specifically, its method of walking and running is referred to as digitigrade, which means that the cat, like dogs and other related carnivores, walks on its toes. This has the effect of adding another segment to the limb, which, in turn, increases length of stride and results in increased speed of running. Man walks on the entire sole of his foot so that the heel touches the ground at each step as well as do the toes. This method of walking is called **plantigrade**.

Before starting the dissection of the hip and thigh region, study the skeletal framework and observe the bones for leverages, the joints for their modifications for stability or range of motion, and the attachments of muscles and ligaments to see how force is exerted. In the hind limb, relate these features to the act of locomotion and compare the adaptations in man and the cat to see both similarities and differences.

Figures 6-1 and 6-2 illustrate the motion of the cat limbs and trunk when the animal is running at top speed. Figure 6-1 shows the position of the limbs and the contact and duration of contact of each limb with respect to the surface of the ground in top-speed running.

Figure 6-2 shows the bony segments of the limbs and the joint positions in the two extremes of trunk extension and trunk flexion.

In the next few laboratory periods, these diagrams should help you understand the action of the musculature of the lower limb, upper limb, and trunk of the cat during rapid locomotion as compared to that of man in bipedal locomotion.

Figure 6-1.

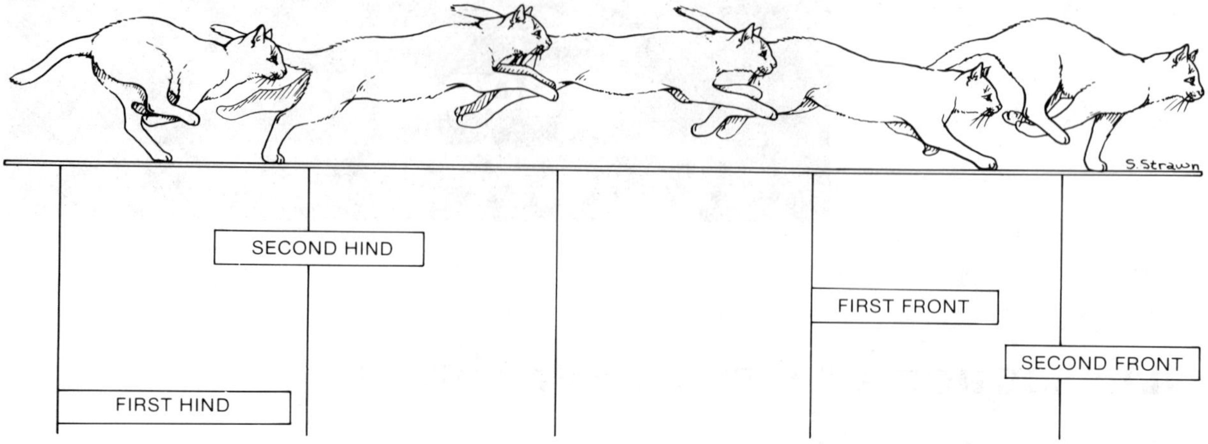

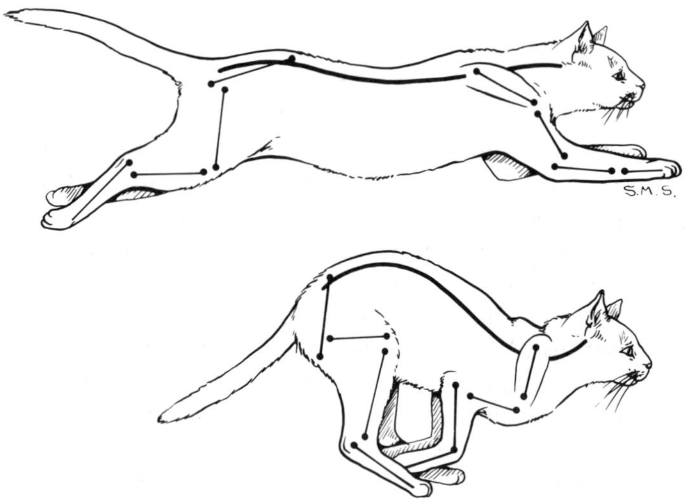

Figure 6-2.

OBJECTIVES

By the end of this laboratory period, you should be able to identify the following: bones, muscles, joints, nerves, and blood vessels of the hip and thigh.

Bones

Ilium
Ischium
Pubis
Femur
Tibia
Fibula

Joints

Sacroiliac
Hip
Knee

Muscles

Hip Flexors
Iliacus
Psoas

Adductors
Adductor magnus
Adductor longus
Adductor brevis
Gracilis
Pectineus

Hip Extensors
Semimembranosus
Semitendinosus
Biceps femoris

Gluteals
Gluteus maximus
Gluteus medius
Gluteus minimus*
Tensor fascia latae

Knee Extensors
Rectus femoris
Vastus lateralis
Vastus medialis
Vastus intermedius
Sartorius

Lateral Rotators*

Nerves

Lumbar Plexus
Femoral nerve
Obturator nerve

Sacral Plexus
Sciatic nerve
Inferior gluteal nerve*
Superior gluteal nerve*

Blood Vessels

Arteries
Femoral artery
Deep femoral artery

Veins
Femoral vein
Deep femoral vein
Great saphenous

*Optional

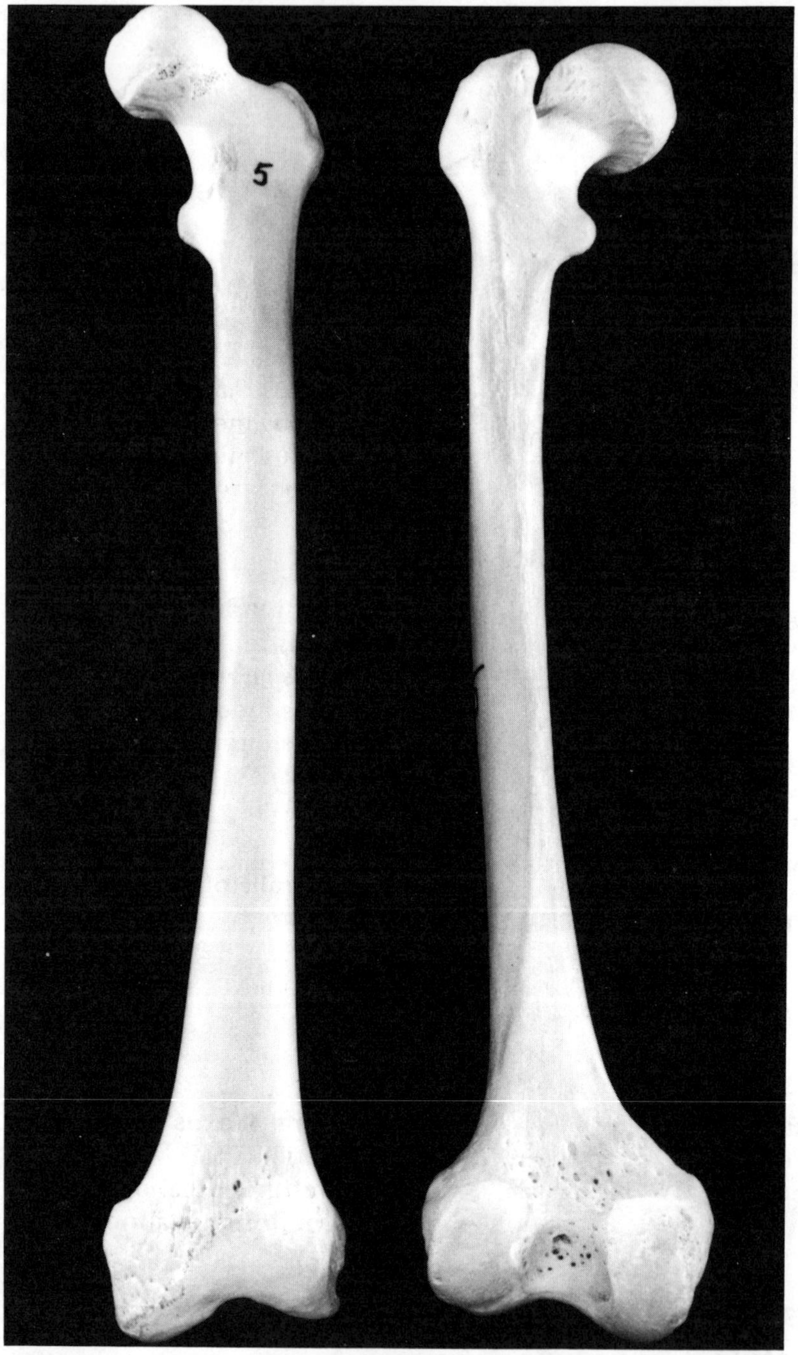

Figure 6-3. Left: ________ aspect of the right femure. Right: ________ aspect of the left femur. Label the following structures with the code indicated. H—head; N—neck; G—greater trochanter; L—lesser trochanter; Ic—Intertrochanteric crest; Il—intertrochanteric line; La-linea alba; Cm—medial condyle; Cl—lateral condyle; Cn—intercondylar notch; P—popliteal fossa. Use this same code to label Figure 6-4.

Figure 6-4.

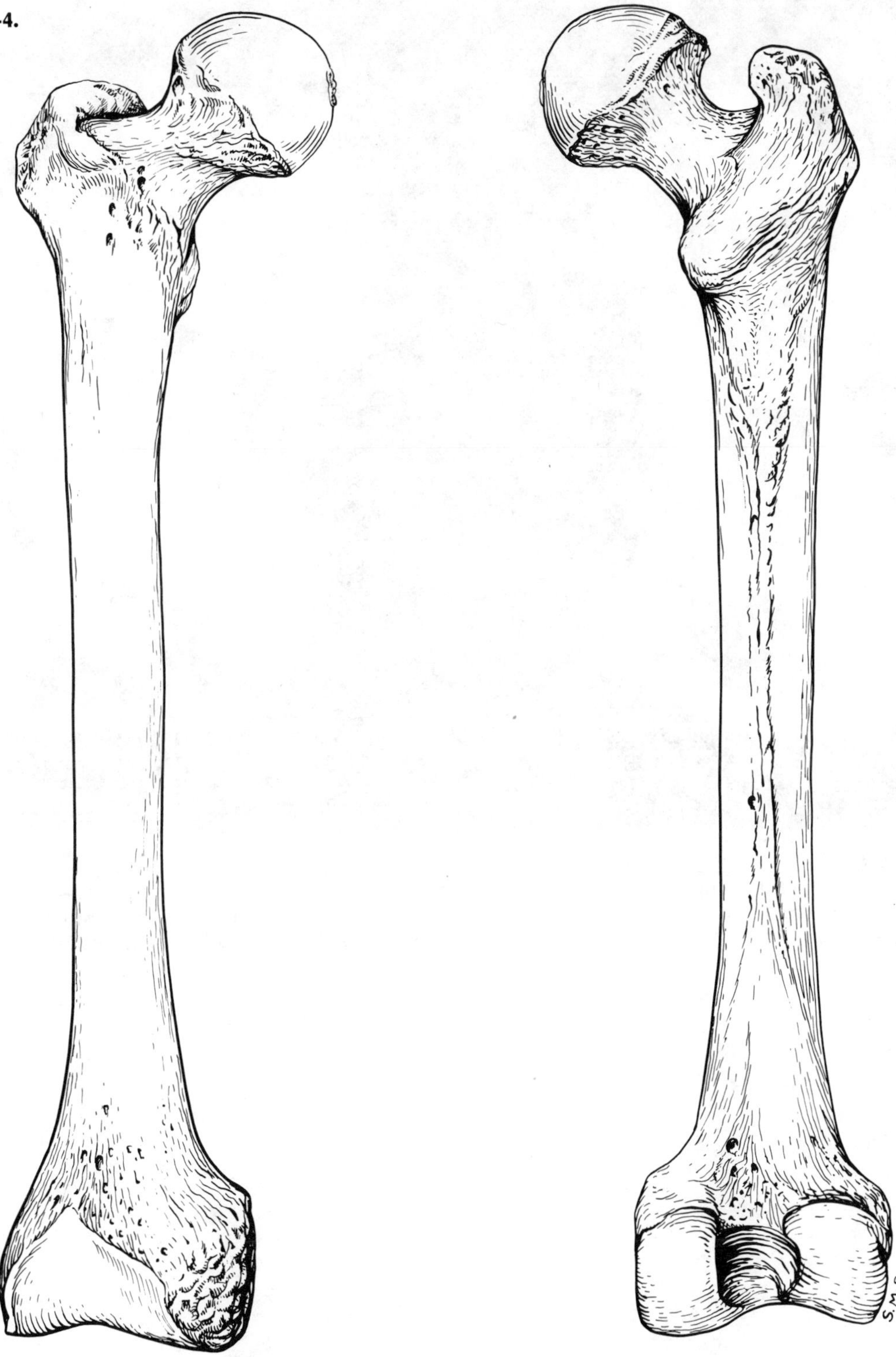

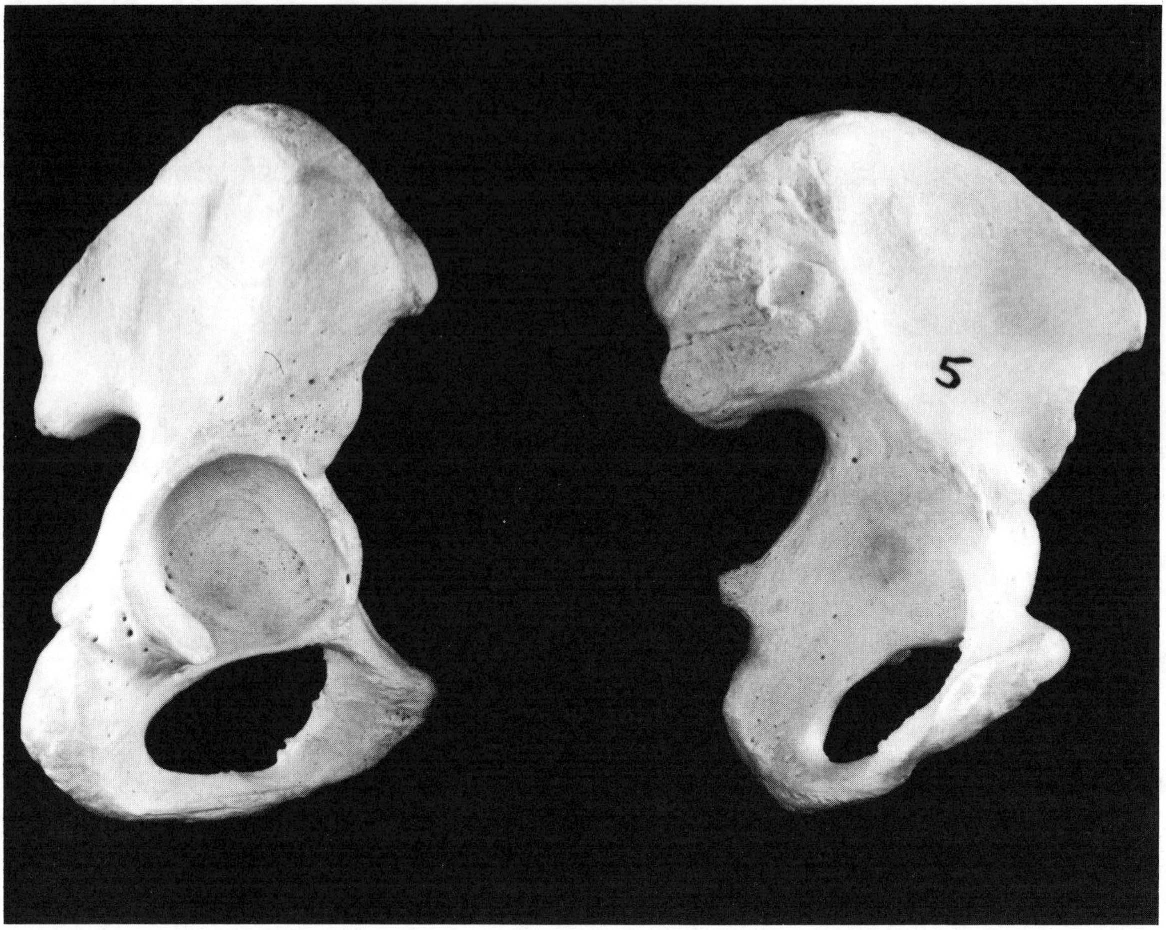

Figure 6-5. Left: ________ aspect of the right pelvic bone. Right: ________ aspect of the left pelvic bone. Label the following structures with the indicated code. I—ilium; P—pubis; Is—Ischium; Au—auricular surface; Ic—iliac crest; It—ischial tuberosity; Ps—pubic symphysis; O—obturator foramen; S—sciatic notch; As—anterior, superior iliac spine. Use this same code to label Figures 6-6 and 6-7.

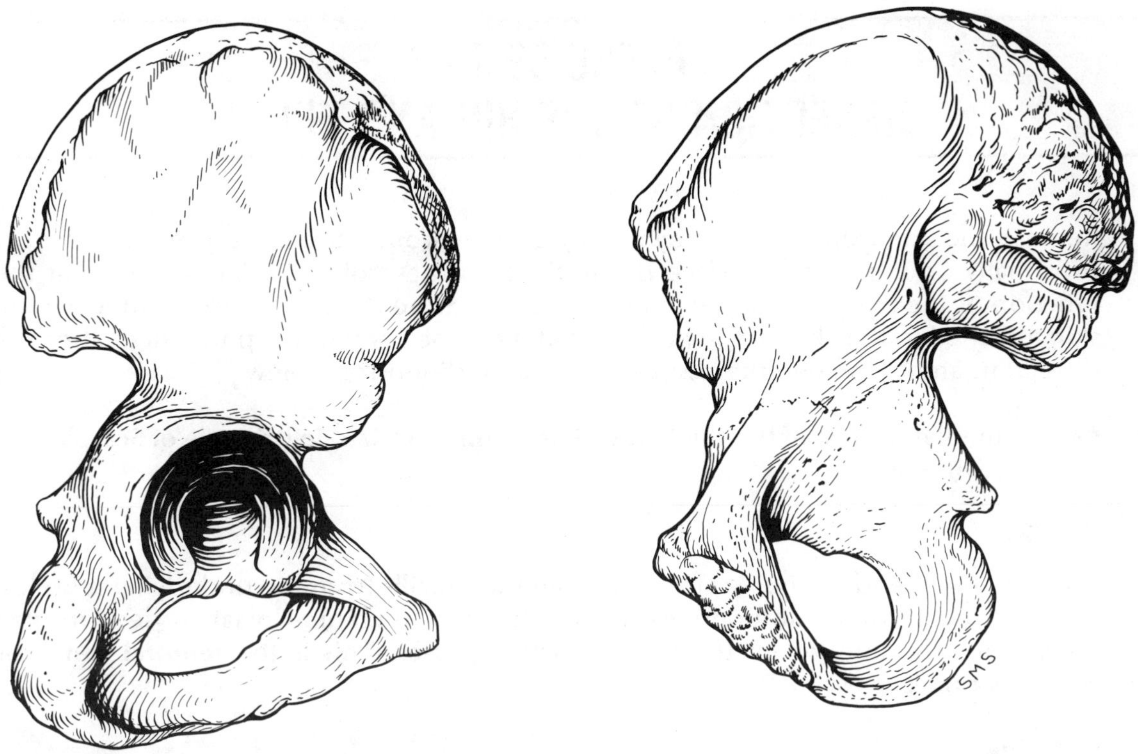

Figure 6-6. Left: lateral view. Right: medial view.

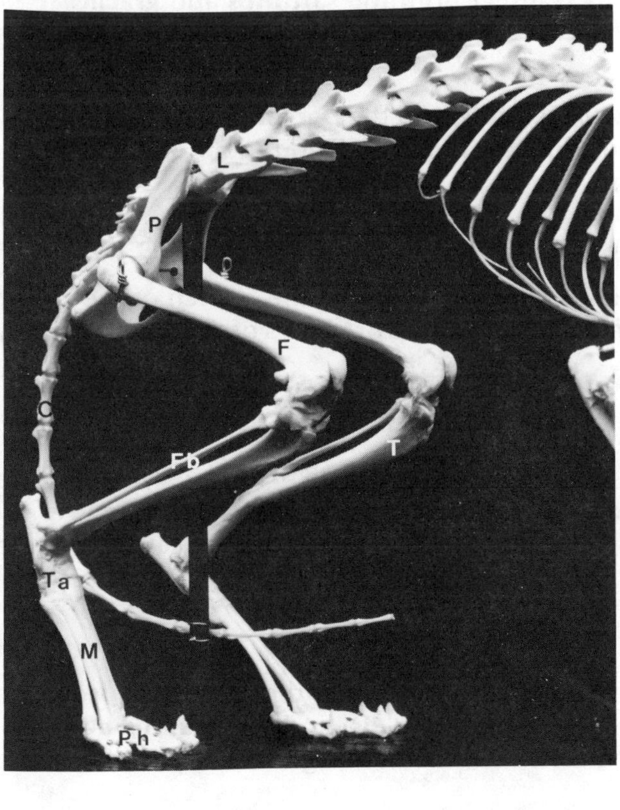

Figure 6-7. Bones of the cat hind limb. P—pelvis; F—femur; T—tibia; Fb—fibula; Ta—tarsals; M—metatarsals; Ph—phalanges; L—lumbar vertebrae; C—caudal vertebrae.

EXERCISE 1
DISSECTION OF THE HIP AND THIGH

1. Remove the skin from the thigh by cutting a circle around the knee. Continue the cut proximally on the medial surface of the thigh to the groin. Cut laterally so that the junction of the thigh and abdomen is exposed. Expose the gluteal area, but leave the skin attached to the skin of the back so that it can be used to wrap the thigh after the dissection and help keep the tissues in good condition for review.

2. On the medial surface of the thigh, locate the **femoral triangle** (triangle of Scarpa) and note its contents.

3. **Gracilis**

   With the blunt end of your scalpel, separate the **gracilis** from the underlying muscles. Note the branch of the **obturator nerve**, which enters its cranio-dorsal border. Observe the groove through which the femoral artery passes. This is the **femoral canal**, or **Hunter's canal.**

4. **Sartorius**

   Dissect the **sartorius.** Notice the branch of the **femoral nerve** that enters the **sartorius** at the ventral border of the proximal end. Transect both the **sartorius** and **gracilis.**

5. **Adductors**

   Dissect the **adductor longus**, the **adductor femoris**, and the **pectineus** (the last is a thin, flat muscle on the dorsal surface of the **adductor femoris**). Find the **obturator nerve** as it emerges from the **obturator foramen.**

6. **Semimembranosus and Semitendinosus**

   Dissect the **semimembranosus** and **semitendinosus.** Observe the branch of the **sciatic nerve** which enters each of these muscles. Note also the branches of the **femoral artery** that supply them.

7. **Tenuissimus and Biceps femoris**

   Locate the **tenuissimus**, which is an extremely thin, straplike muscle lying on the interior surface of the **biceps femoris** and covering the **sciatic nerve.** This muscle is not ordinarily present in man. Dissect and transect the **biceps femoris.** Locate the nerve and blood supply of this muscle.

8. **Gluteals**

   Find the **tensor fascia lata.** Transect its tendon of insertion and dissect away from the **gluteus maximus** and **medius.** Locate the **caudofemoralis**, and the **gluteus minimus.** The **inferior gluteal** nerve may be traced back from the **gluteus maximus** or the

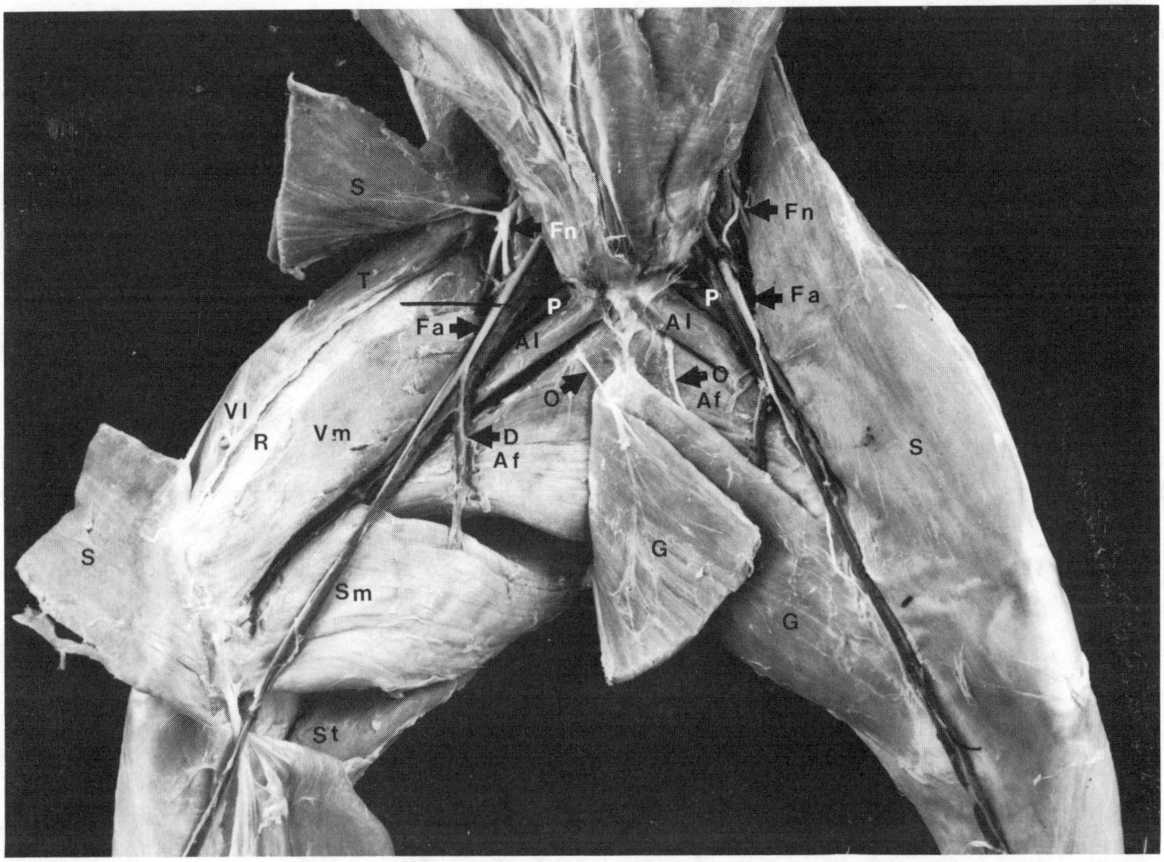

Figure 6-8. Dissection of hip and thigh, medial aspect. Muscles: G—gracilis; S—sartorius; Al—adductor longus; Af—adductor femoris; Sm—semimembranosus; St—semitendinosus; P—pectineus; Vl—vastus lateralis; Vm—vastus medialis; R—rectus femoris; T—tensor fascia lata. Nerves: O—obturator nerve; Fn—femoral nerve. Blood Vessels: Fa—femoral artery; D—deep femoral artery.

caudofemoralis (not present in man). Likewise, the **superior gluteal** nerve may be seen emerging from under the lateral border of the **piriformis**.

9. **Quadriceps**

On the front of the thigh, observe the **quadriceps femoris**, which is made up of the **vastus lateralis**, the **vastus medialis**, the **rectus femoris**, and the **vastus intermedius**. The last of these will not be visible until the **rectus femoris** is transected. What is the nerve and blood supply of these muscles?

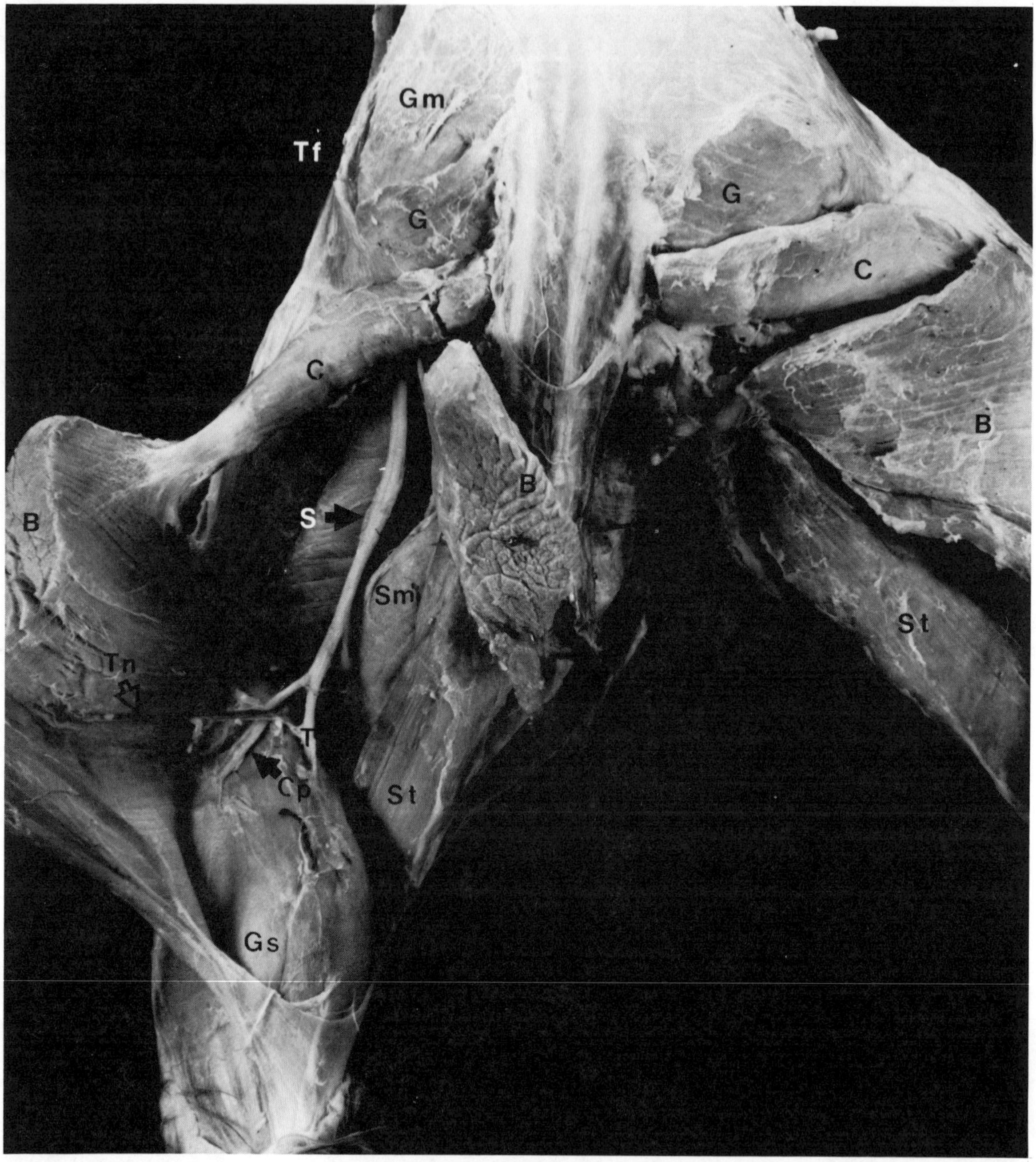

Figure 6-9. Dissection of hip and thigh, lateral and posterior aspects. Muscles: B—biceps femoris; Sm—semimembranosus; St—semitendinosus; Tn—tenuissimus; Tf—tensor fascia lata; G—gluteus maximus; Gm—gluteus medius; Gs—gastrocnemius (lateral head); C—caudofemoralis. Nerves: S—sciatic nerve; T—tibial nerve; Cp—common peroneal nerve.

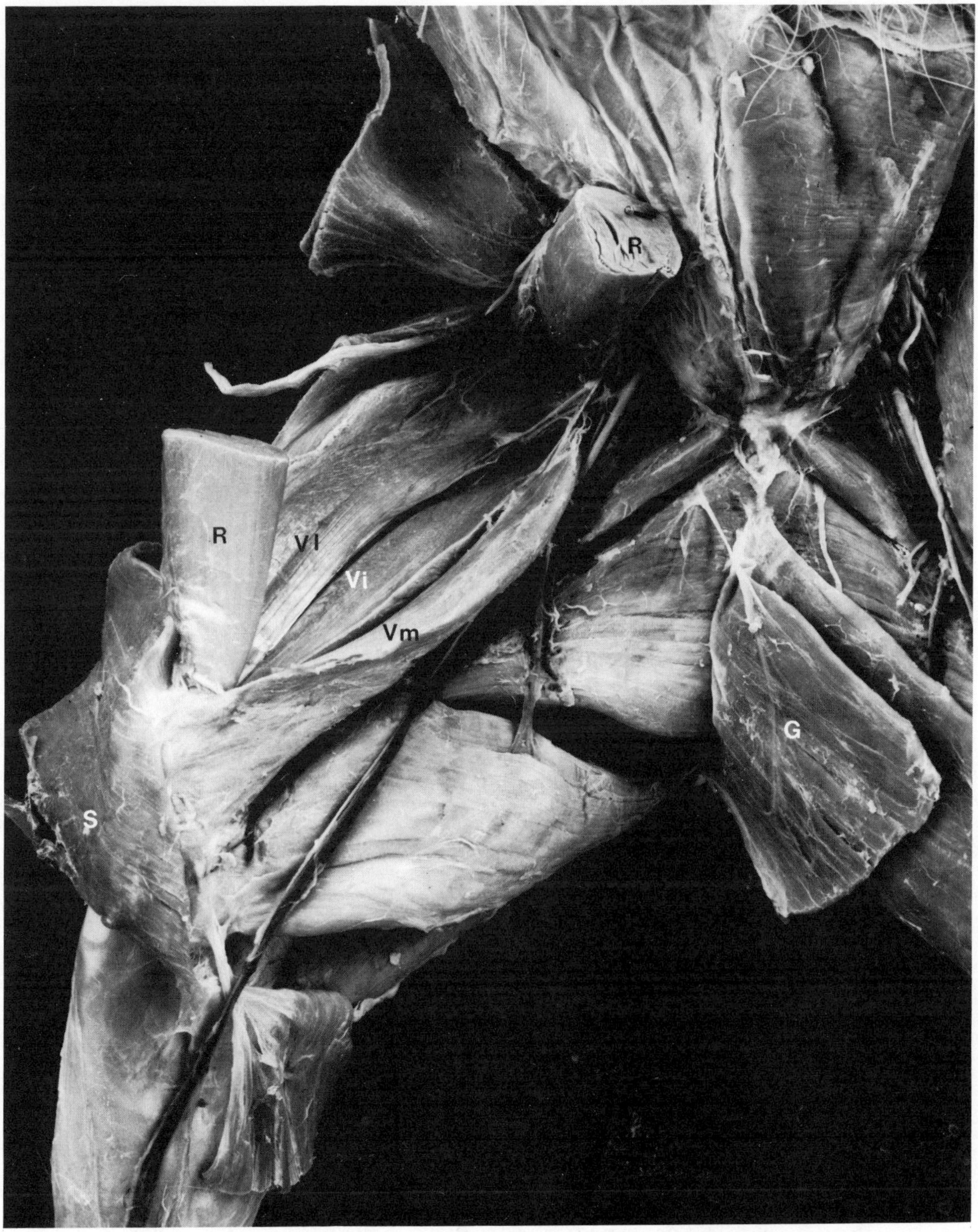

Figure 6-10. Dissection of quadriceps, medial aspect. Muscles: G—gracilis; S—sartorius; R—rectus femoris; Vm—vastus medialis; Vl—vastus lateralis; Vi—vastus intermedius.

Figure 6-11. Dorsal muscles.

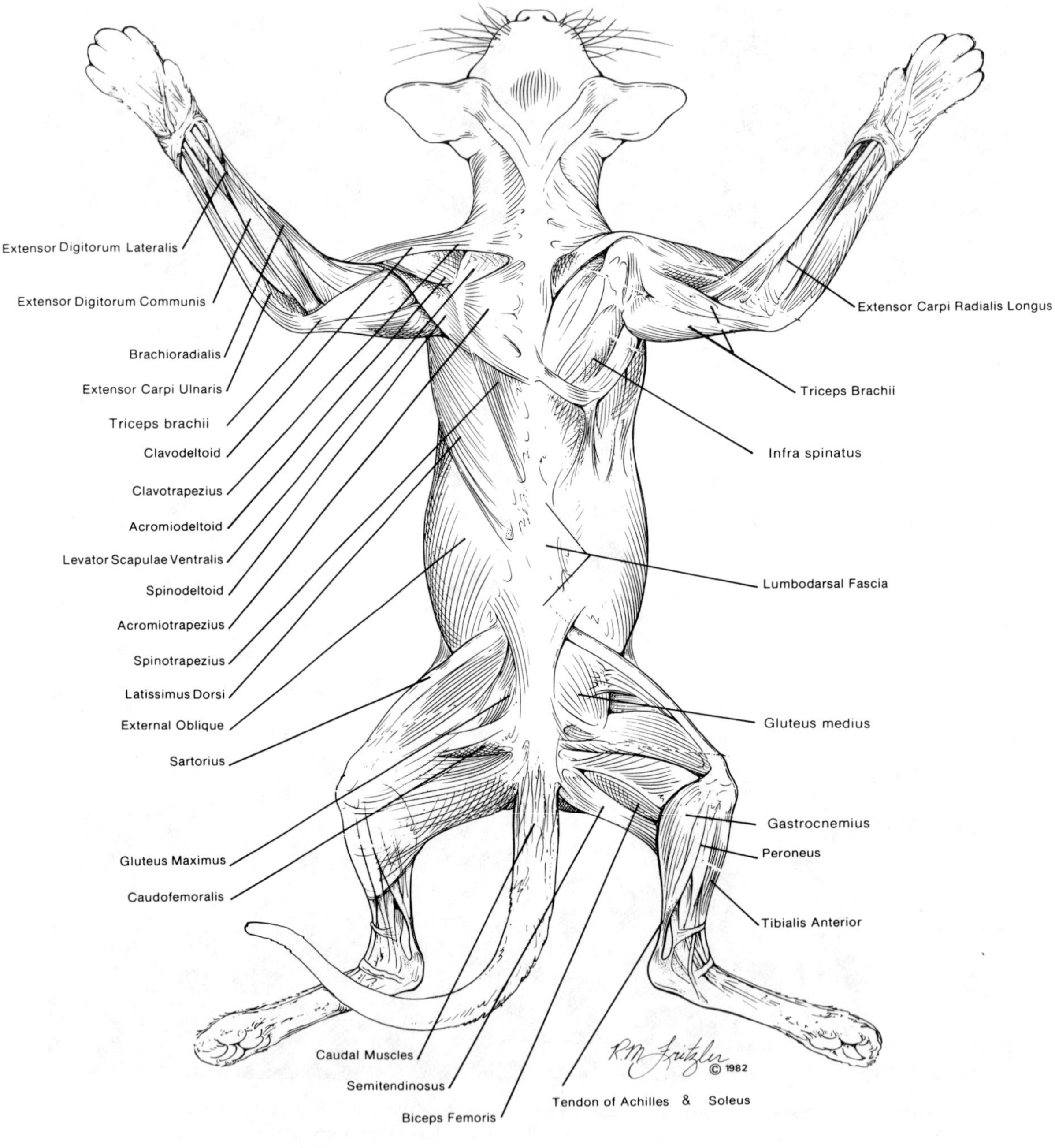

Figure 6-12. Ventral muscles.

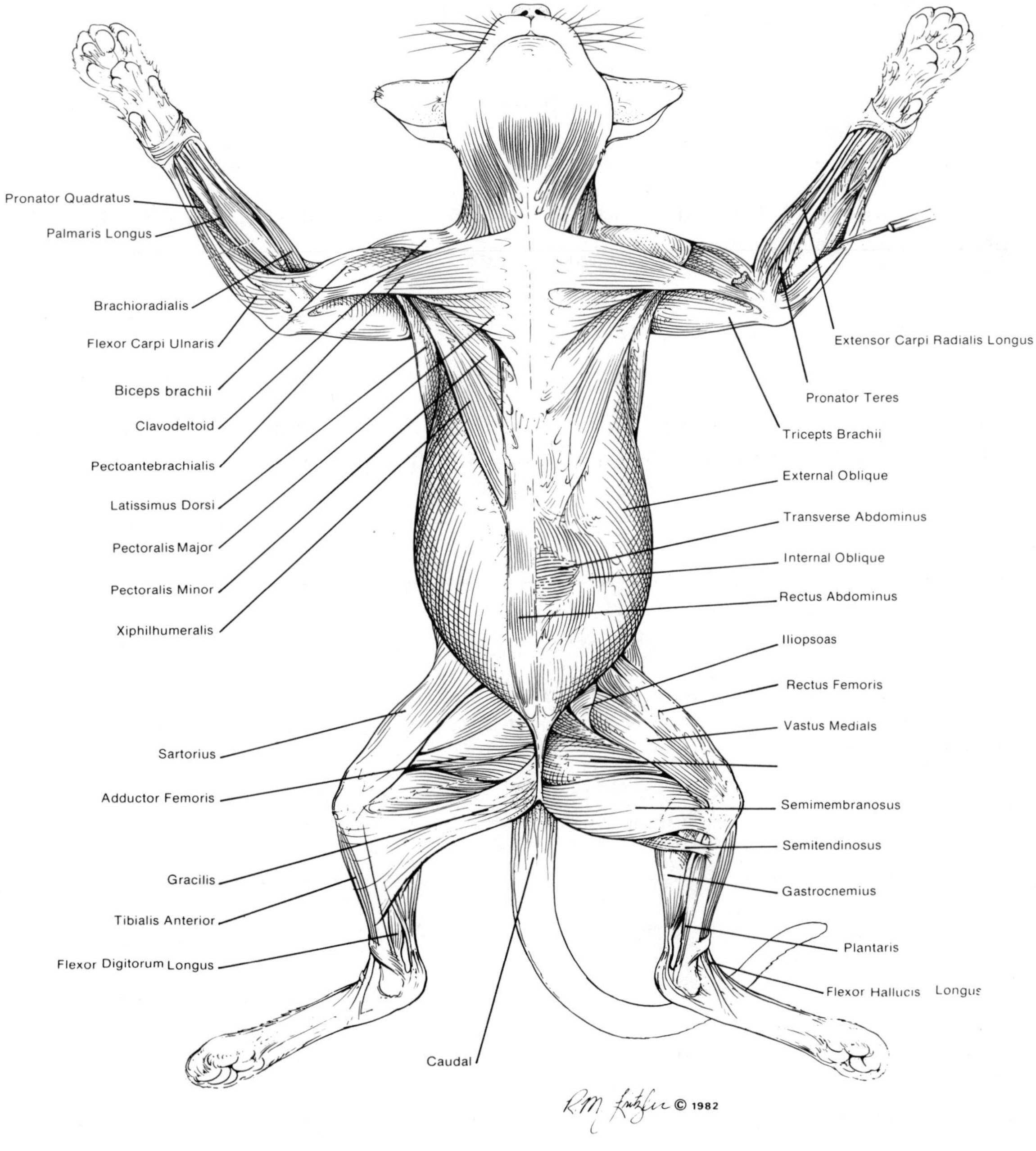

EXERCISE 2
DISCUSSION QUESTIONS

1. What features of the pelvic girdle of man are adaptations for erect posture and bipedal locomotion?

2. How does the hip joint of the cat differ from that of man in structure, stability, and provision for range of motion?

3. What is the principal difference in hip musculature between man and the cat? What characteristic of human motion is made possible by this difference?

4. What muscles are the "power" extensors of the hip joint in the cat? Does this help explain the advantages of a crouching start in sprinting and the high seat position in bicycle racing?

5. What are contents of the **femoral triangle**? What veins are joined together there?

LABORATORY 7

Dissection of Cat Lower Limb II: Shank and Foot

OBJECTIVES

By the end of this laboratory, you should be able to identify the following: bones, joints, muscles, nerves, and blood vessels of the shank and foot.

Bones

Tibia
Fibula
Patella
Tarsals
Metatarsals
Phalanges

Joints

Knee
Ankle
Subtalar
Midtarsal joint
Composite joint
Tarso-metatarsal joint

Muscles

Dorsiflexors
Tibialis anterior
Extensor digitorum longus

Evertors
Peroneus longus
Peroneus tertius
Peroneus brevis

Plantar Flexors
Gastrocnemius
Soleus
Plantaris
Tibialis posterior

Digital Flexors and Plantar Flexors
Flexor digitorum longus, first head
Flexor digitorum longus, second head

Nerves

Tibial nerve
Common peroneal nerve
Deep peroneal nerve
Superficial peroneal nerve

Blood Vessels

Arteries
Popliteal artery
Posterior tibial artery
Anterior tibial artery
Posterior geniculate artery*
Suralis*

Veins
Dorsal Venous arch
Short saphenous vein
Long saphenous vein
Popliteal vein
Anterior tibial vein

*Optional

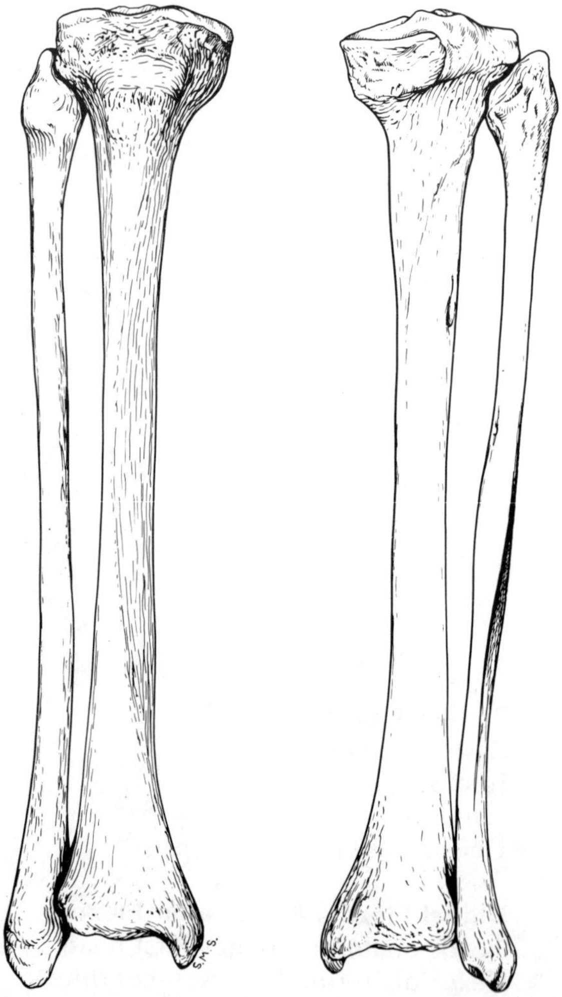

Figure 7-1. Left: right tibia and fibula from in front. Right: right tibia and fibula from behind.

Label the following structures as indicated. T—tibial condyle; C—crest; Tt—tibial tuberosity; L—lateral malleolus; M—medial malleolus; H—head of fibula. Use the same code to label Figure 7-1.

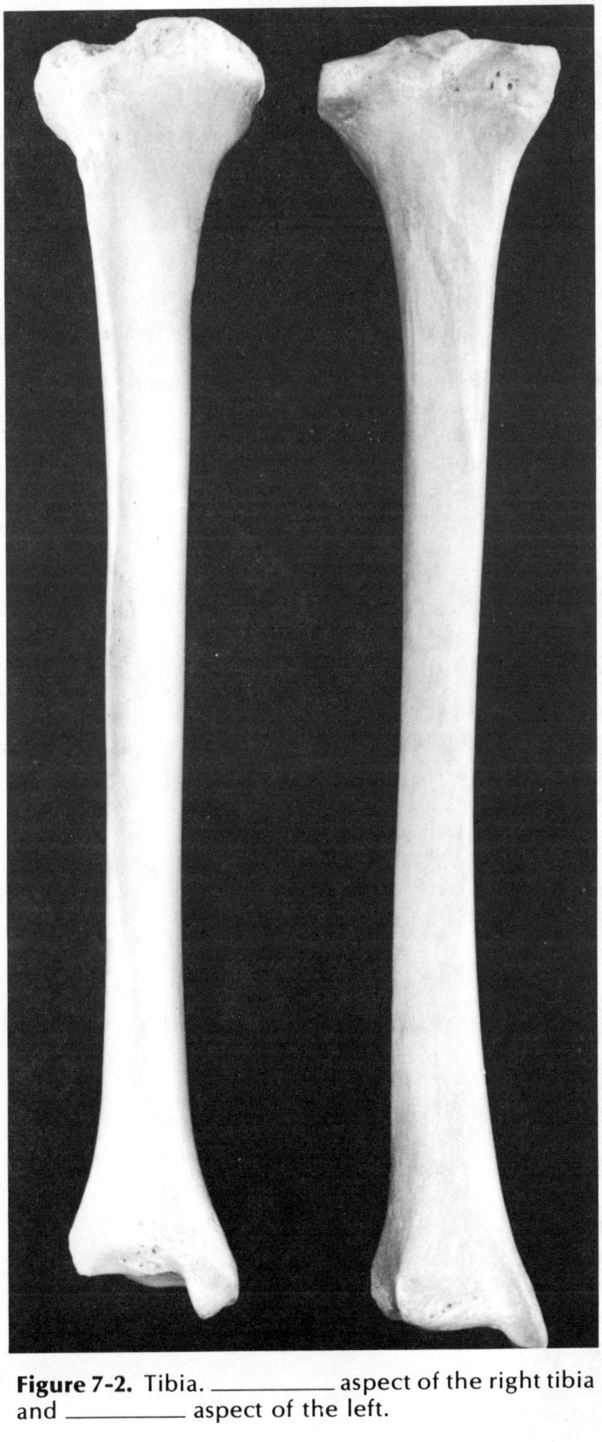

Figure 7-2. Tibia. __________ aspect of the right tibia and __________ aspect of the left.

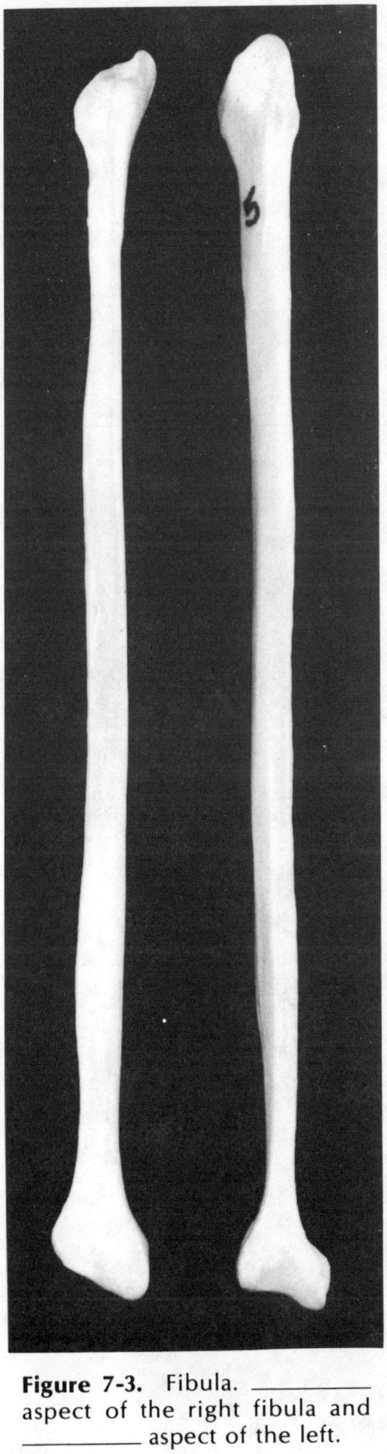

Figure 7-3. Fibula. __________ aspect of the right fibula and __________ aspect of the left.

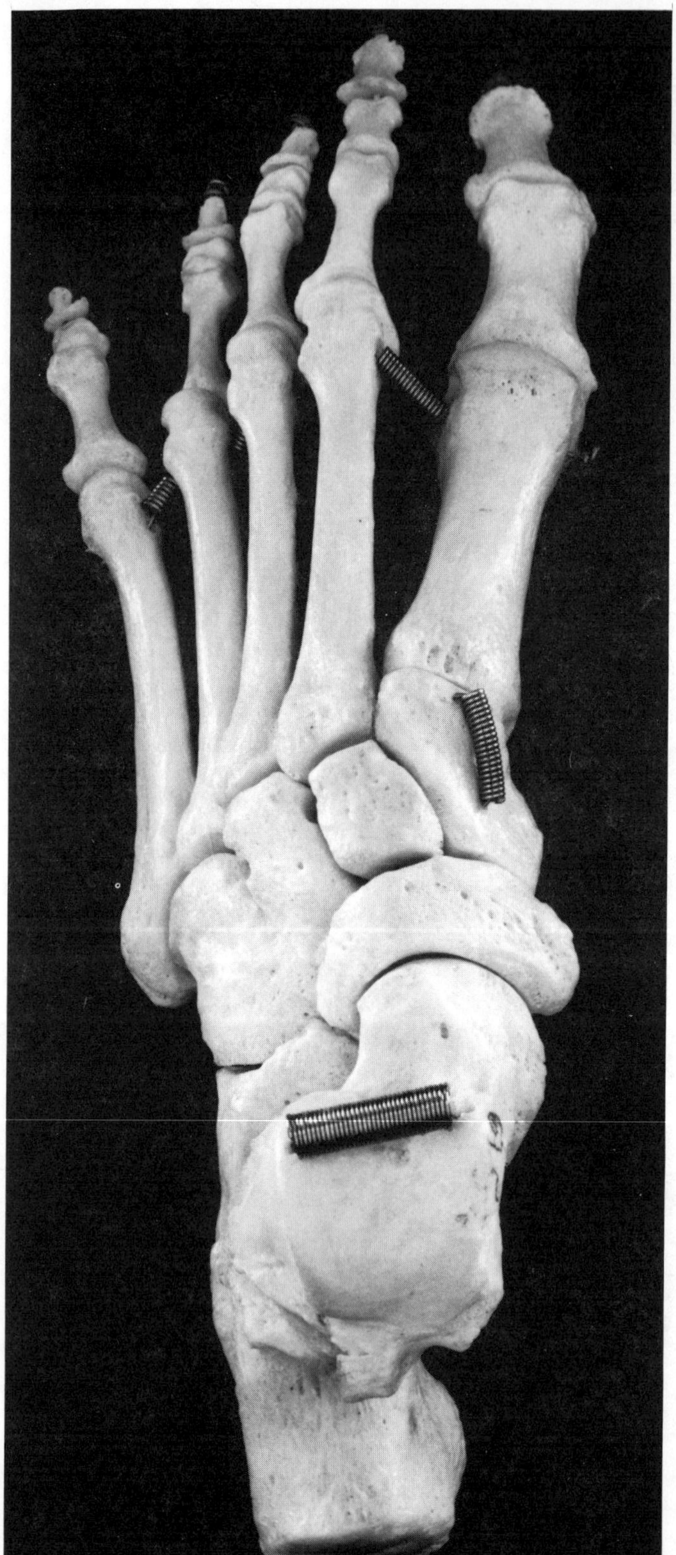

Figure 7-4. Left foot, dorsal view. Label the bones in figure 7-4 as follows: C—calcaneus; T—talus; N—navicular; Cu—cuboid; C1—first cuneiform; C2—2nd cuneiform; C3—3rd cuneiform; M1-M5—metatarsals; P—phalanges. Use this same code to label Figures 7-5, 7-6, 7-7, and 7-8.

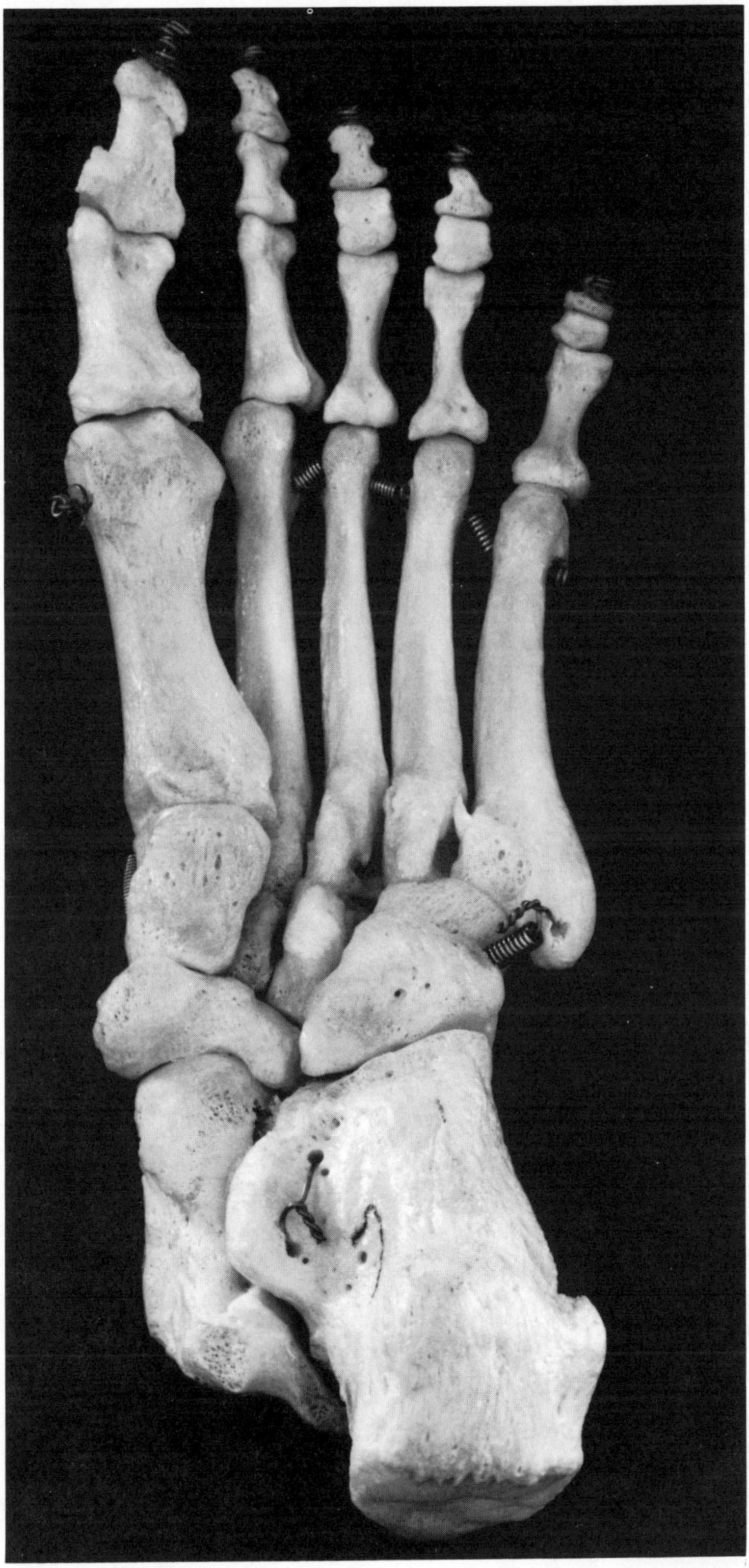

Figure 7-5. Left foot, plantar view. Note spur at base of 5th metatarsal.

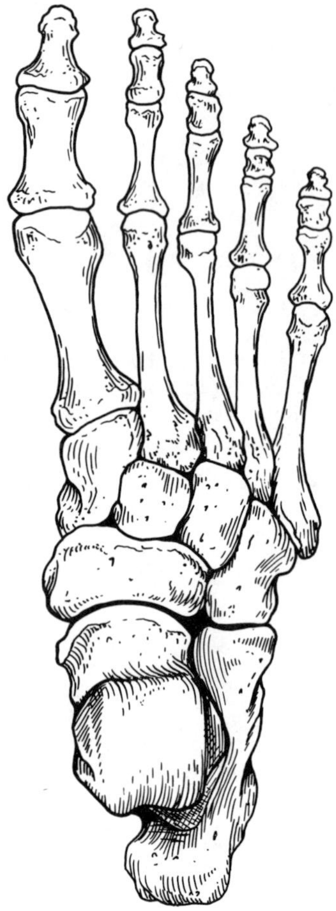

Figure 7-6. Bones of right foot viewed from above.

Figure 7-7. Bones of right foot, medial view.

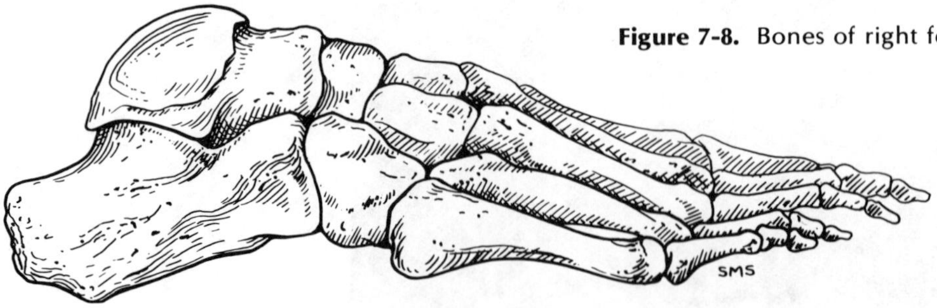

Figure 7-8. Bones of right foot, lateral view.

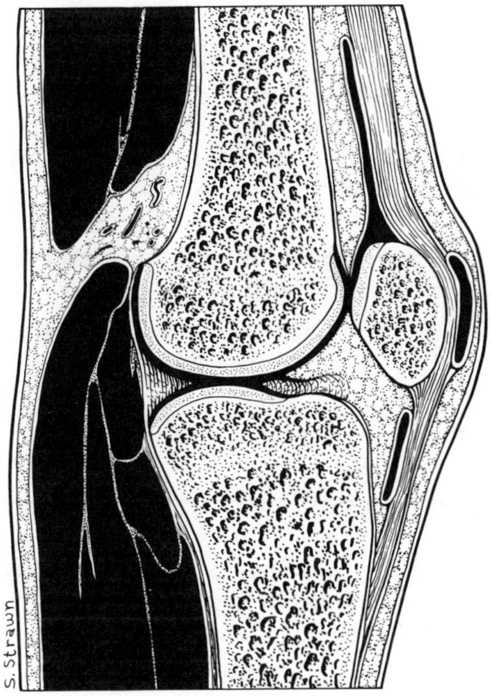

Figure 7-9. Dissection of knee joint; sagittal section through lateral condyles of femur and tibia.

Label the following:
1. Femur
2. Tibia
3. Patella
4. Patellar ligament
5. Articular surfaces
6. Bursae
7. Adipose tissue

EXERCISE 1
DISSECTION OF THE SHANK AND FOOT

1. Remove the skin from the entire shank and foot, exposing the musculature, tendons, and fascia.

2. **Plantar flexors**

   On the posterior surface of the shank, notice the **gastrocnemius**, a large muscle with two heads fused together in the central region but split near the origin and insertion on the **Achilles tendon**. The two main divisions of the **sciatic nerve** separate from each other in the popliteal space, which is partially formed by the cleft between the heads of the **gastrocnemius**. The medial branch is the **tibial nerve**, and it passes between the heads of the **gastrocnemius**, giving off branches to the **plantaris** and **soleus** and finally descending to the plantar surface of the foot.

   Cut the **tendon of Achilles** and reflect back the muscle mass. The heads of the **gastrocnemius** enclose both the **plantaris** and the **soleus**. The **plantaris** originates on the **patella** and inserts on the **flexor digitorum brevis** (on the sole of the foot) after passing over the **calcaneus** by a tubular sheath. The **soleus** is deep to the **plantaris** and originates from the head and upper ventral border of the **fibula**. Identify these muscles and their innervation and blood supply.

3. **Dorsiflexors**

Dissect the muscles of the anterior surface of the shank, and carefully separate their tendons of insertion. In order to do this, the ringlike **annular ligament** of the ankle must be cut. Identification of the muscles is easily accomplished after complete and thorough dissection of the tendons of insertion. The **tibialis anterior** is the largest and most conspicuous muscle of the front of the shank. It originates from the proximal region of the **tibia** and **fibula** and inserts by way of a short, thick tendon of insertion on the medial surface of the first **metatarsal**. The **extensor digitorum longus (EDL)** lies immediately lateral to the **tibialis anterior**. It can be easily identified by its four long tendons of insertion which split from the muscle just after passing through the **annular ligament** of the ankle and insert on the terminal phalanges. Identify these muscles and their nerve and blood supply.

4. **Peroneals**

The peroneal group occupies most of the lateral surface of the shank. The largest of these is the **peroneus longus**, which is in a conspicuous position between the **tibialis anterior** and the lateral head of the **gastrocnemius**. It partially lies over the **extensor digitorum longus**. The **peroneus tertius** is a long, slender muscle lying just mediad to the **peroneus longus**, and the **peroneus brevis** is a short muscle lying mediad to the **peroneus tertius**. Locate these muscles and trace the route of their tendons of insertion around the **lateral maleolus**. Find their nerve and blood supply.

5. **Digital flexors and Plantar flexors**

On the medial surface of the shank, the **popliteus** and the **flexor digitorum longus** can be located on the ventral surface of the **tibia**. The **popliteus** inserts on the ventromedial surface of the proximal end of the **tibia**, while the **FDL** originates there and may be found just under the distal border of the **popliteus**. The **flexor hallucis longus** (more properly termed the second head of the **flexor digitorum longus**) is a long, thick muscle lateral to the **FDL**. These two muscles have a common tendon of insertion which splits into four divisions and inserts on the bases of the terminal phalanges. The **tibialis posterior** is a narrow muscle lying lateral to the **FDL** and medial to the **FHL**. Locate the nerve and blood supply of these muscles.

6. **Nerves**

After splitting of the **sciatic nerve** into the **tibial** and **common peroneal** nerves, the latter passes between the **EDL** and the **peroneus longus**, where it divides into the **deep peroneal** and the **superficial peroneal**. The **deep peroneal** gives off branches to the **tibialis anterior** and **EDL** and runs into the medial side of the dorsal surface of the foot. The **superficial peroneal** gives off branches to the peroneal muscles and also passes into the dorsal surface of the foot.

7. **Blood vessels**

The **femoral artery** becomes the **popliteal artery** when it passes through the **popliteal fossa**. The **popliteal artery** branches into: (a) the **suralis**; (b) muscular branches; (c) posterior **geniculate**; (d) posterior **tibial**; and (e) the anterior **tibial**. The **suralis** passes

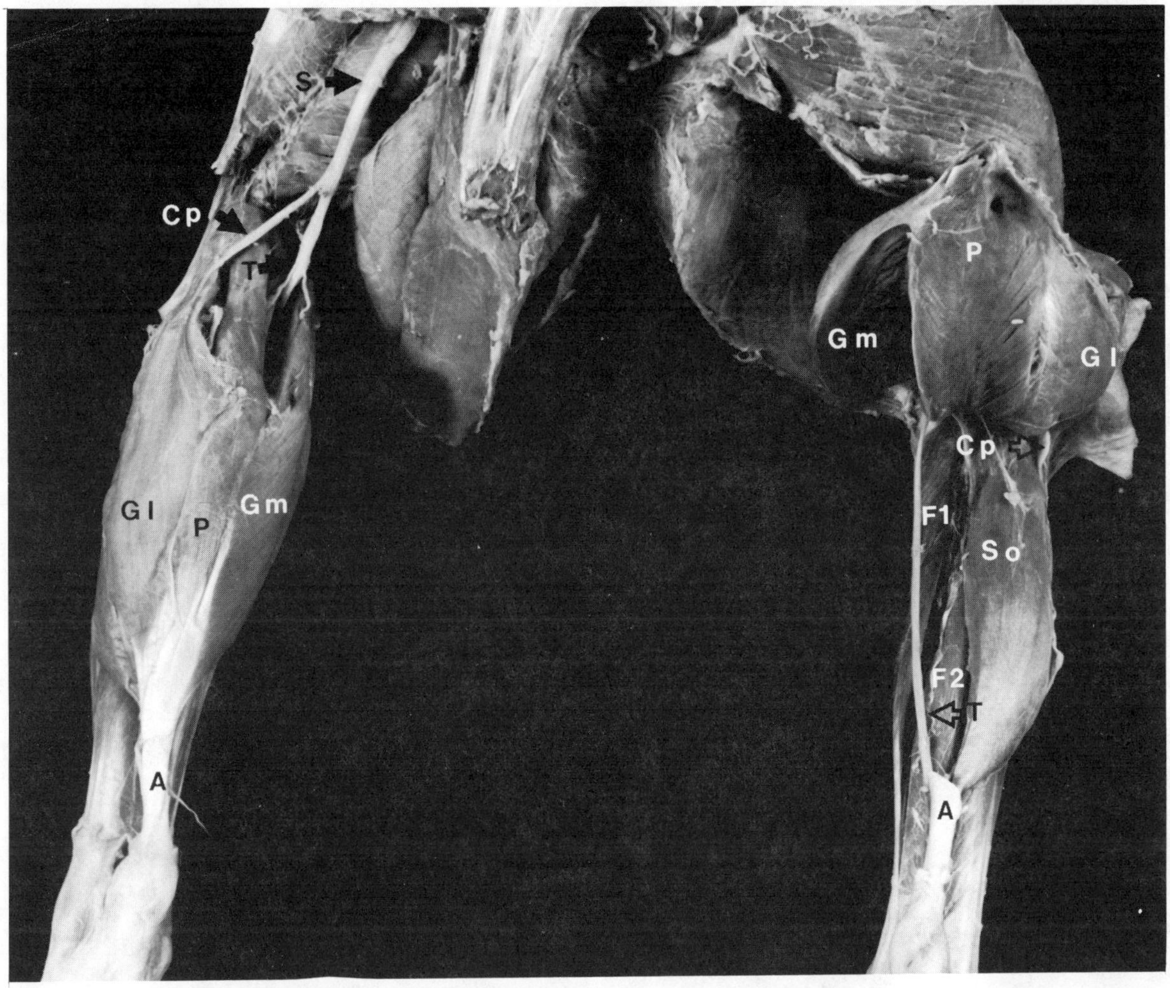

Figure 7-10. Dissection of cat shank, posterior aspect. Muscles: Gl—lateral head of gastrocnemius; Gm—medial head of gastrocnemius; P—plantaris; So—soleus; A—Achilles tendon; F1—flexor digitorum longus, first head; F2—flexor digitorum longus, second head. Nerves: S—sciatic; T—tibial; Cp—common peroneal.

over the surface of the lateral head of the **gastrocnemius**. The muscular branches split off to the musculature in the vicinity of the **popliteal fossa**. The posterior **geniculate** ramifies on the medial surface of the knee. The posterior **tibial** runs down along the medial surface of the **flexor hallucis longus** along with the **tibial nerve**. The anterior **tibial artery** penetrates the interosseus membrane between the **tibia** and **fibula** and then descends distally between the **tibialis anterior** and the **EDL**.

The deep veins of the cat hind limb run along with the above named arteries. The superficial veins are the short **saphenous** on the lateral surface of the shank and the long **saphenous** on the medial surface.

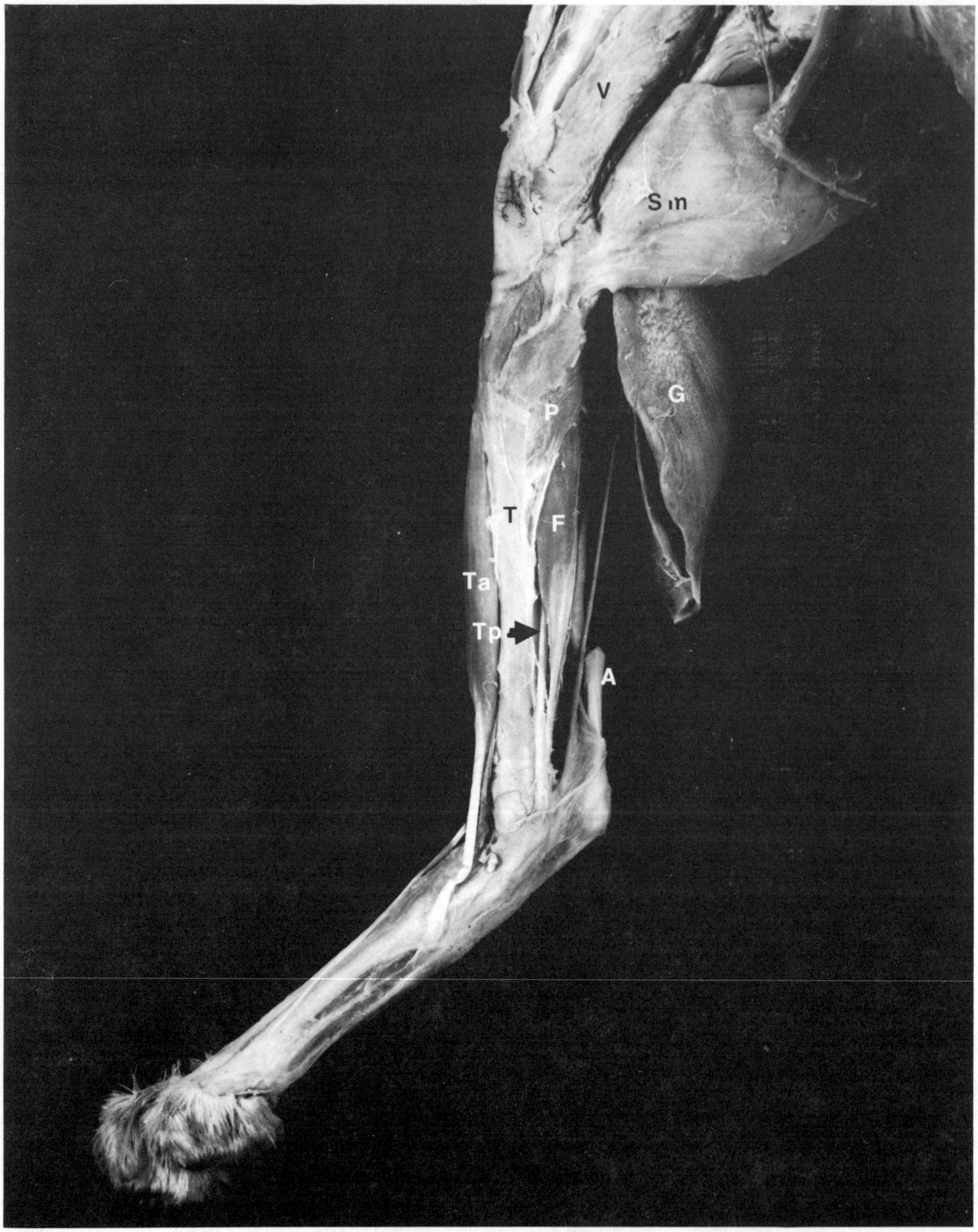

Figure 7-11. Dissection of cat shank, medial aspect. C—vastus medialis; Sm—semimembranosus; G—medial head of gastrocnemius; P—popliteus; F—flexor digitorum longus, first head; Tp—tendon of tibialis posterior; Ta—tibialis anterior; A—Achilles tendon; T—tibia.

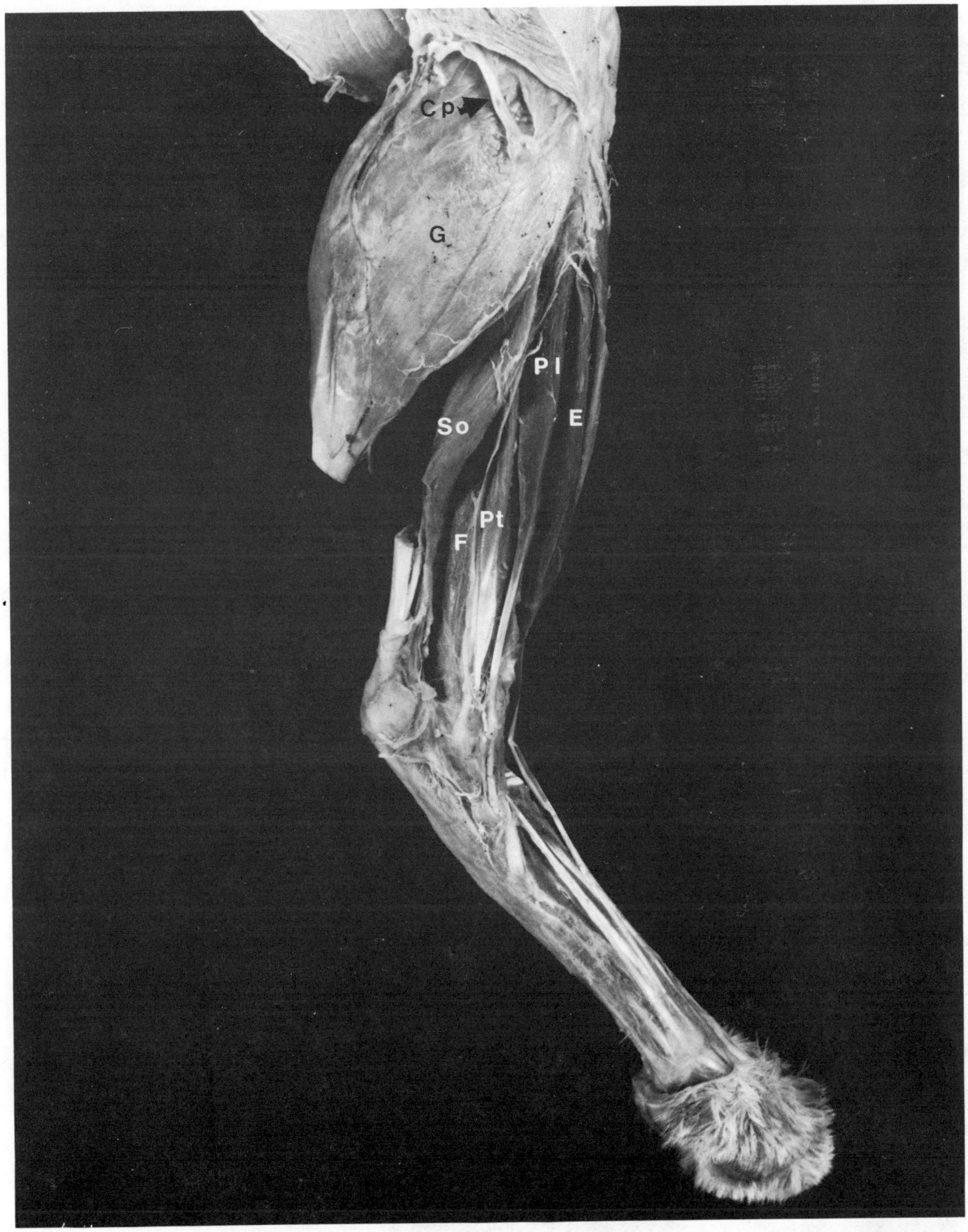

Figure 7-12. Dissection of cat shank, lateral view. Muscles: G—lateral head of gastrocnemius; So—soleus; F—flexor digitorum longus, second head; Pl—peroneus longus; Pt—peroneus teritius; E-extensor digitorum longus. Nerves: Cp—common peroneal.

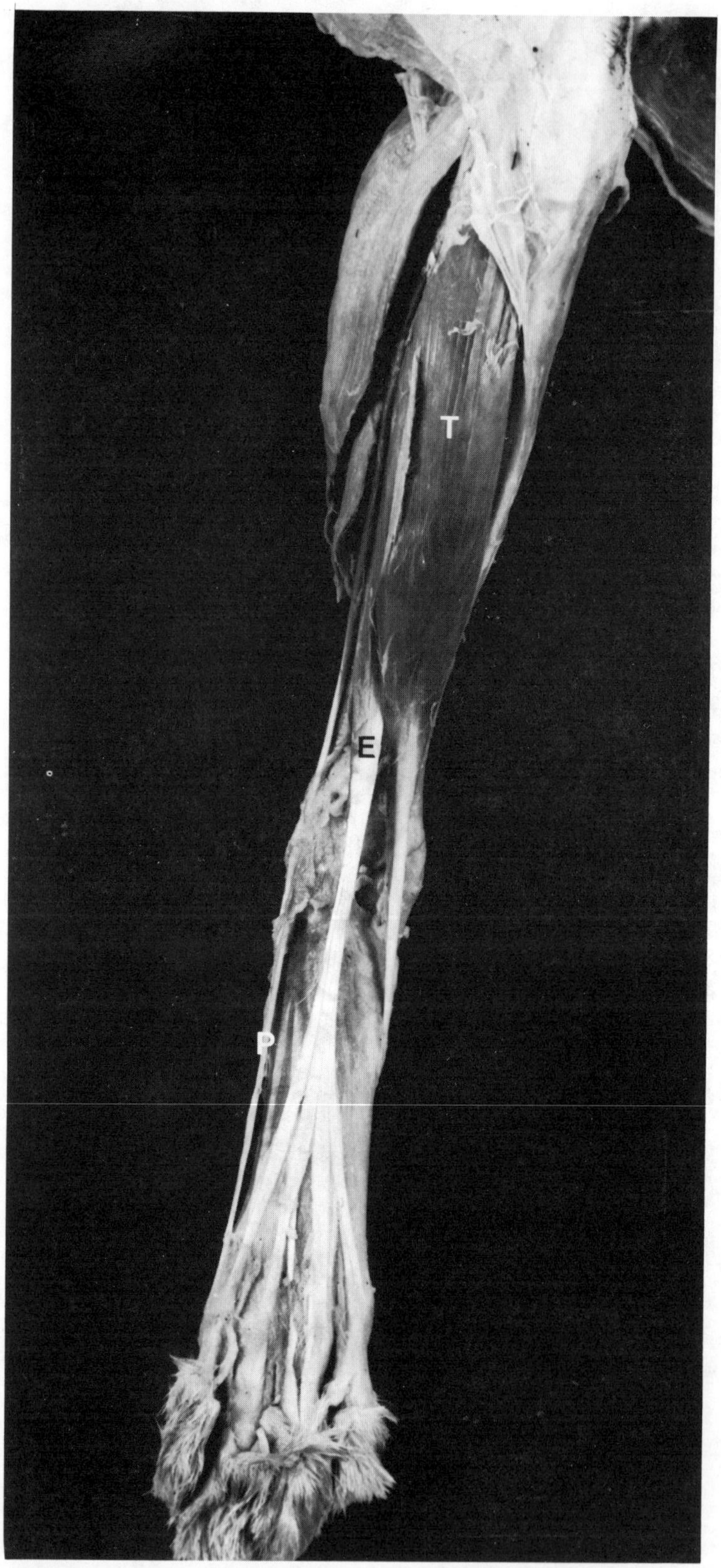

Figure 7-13. Dissection of cat shank, anterior view. T—tibialis anterior; E—extensor digitorum longus; P—tendon of peroneus longus.

EXERCISE 2
DISCUSSION QUESTIONS

1. Compare the plantar flexors of the cat with those of man. What is the explanation for this difference?

2. What muscle of the front compartment of the shank is present in man but not in the cat? What is its function in man?

3. Explain the specialization of the foot in man and compare it with the foot of the cat with regard to relative size of bones and attachments of muscles and ligaments. What is the significance of the relatively long metatarsals in the cat?

4. How would the cat's gait be altered if the femoral nerve were severed at the junction of the thigh and abdomen?

5. What muscles would be paralyzed if the sciatic nerve were injured at the point of its division above the popliteal space?

____

____

6. How would the cat's gait be affected by the above injury? How would it affect the gait in man?

____

____

____

____

LABORATORY 8

Dissection of Cat Forelimb I: Shoulder, Upper Arm, and Pectoral Region

OBJECTIVES

By the end of this laboratory period, you should be able to identify the following: bones, muscles, nerves, and blood vessels of the shoulder and upper arm.

Bones

Vertebrae
Clavicle
Scapula
Humerus

Muscles

Trapezii
Spinotrapezius
Acromiotrapezius
Clavotrapezius

Muscles Acting on the Scapula
Levator scapulae ventralis
Rhomboideus
Latissimus dorsi
Serratus anterior
Rhomboideus capitus

Deltoids
Spinodeltoideus
Acromiodeltoideus
Clavodeltoideus

Muscles of the Scapula
Subscapularis
Supraspinatus
Infraspinatus
Teres major
Teres minor

Pectorals
Pectoantebrachialis
Pectoralis major
Pectoralis minor
Xiphihumeralis

Muscles of the Brachial Region
Biceps brachii
Triceps brachii
Brachialis
Anconeus
Coracobrachialis

Blood Vessels

Subclavian artery and vein
Axillary artery and vein

Nerves

Spinal accessory
Axillary
Musculocutaneous
Radial

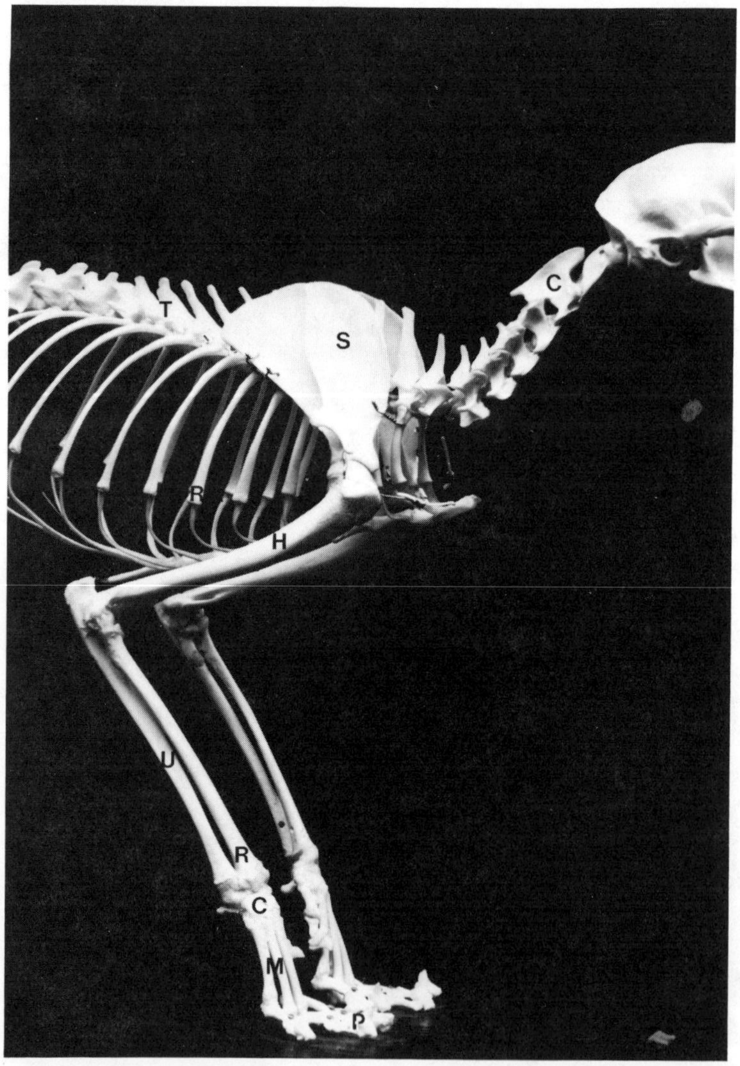

Figure 8-1. Bones of cat forelimb. S—scapula; H—humerus; R—radius; U—ulna; C—carpals; M—metacarpals; P—phalanges; T—thoracic vertebrae; C—cervical vertebrae; R—ribs.

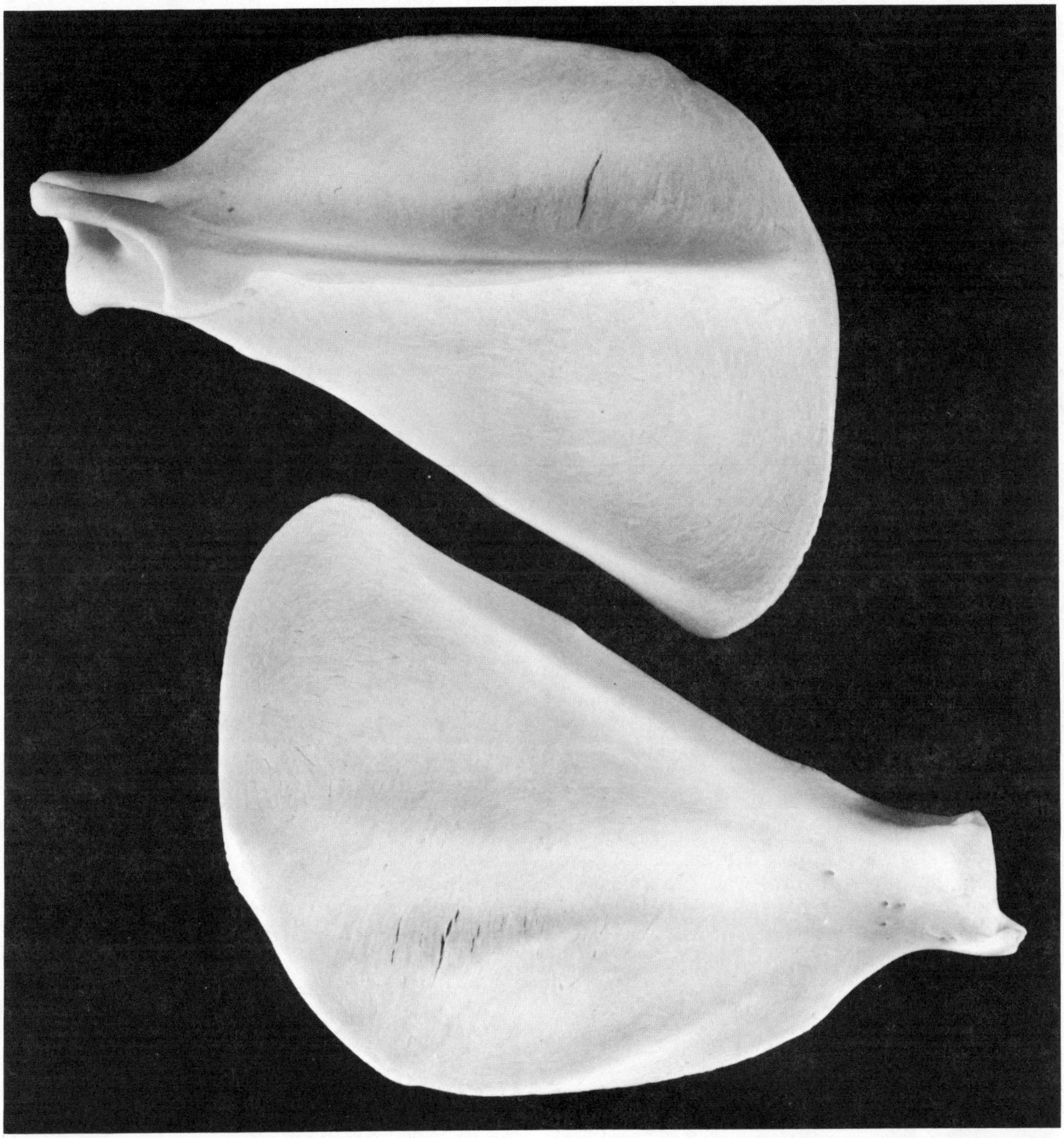

Figure 8-2. Cat scapula; top—dorsal view; bottom—costal view. How does the cat scapula differ from that of man? Why?

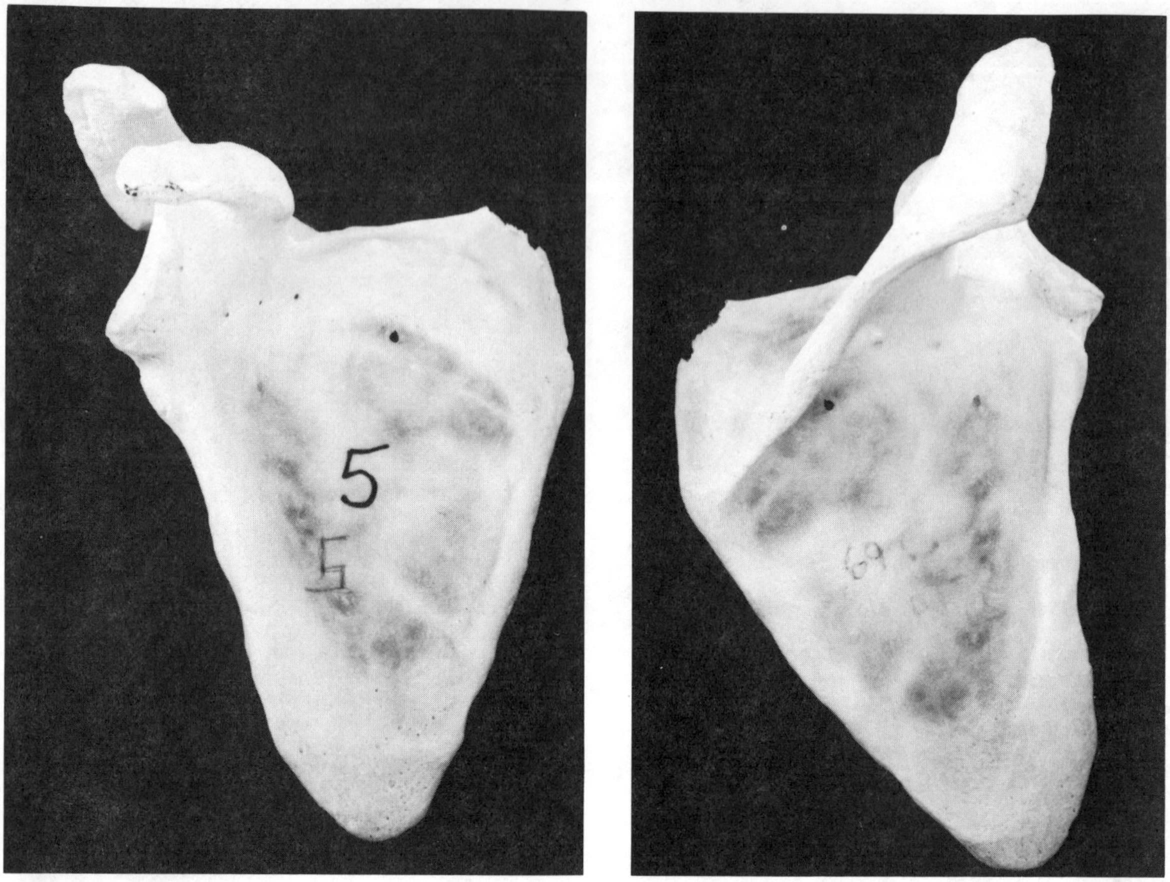

Figure 8-3. Left: costal aspect of the __________ scapula. Right: dorsal aspect of the __________ scapula. Label the following structures with the indicated code. Su—subscapular fossa; G—glenoid fossa; S—spine; Ss—supraspinous fossa; Is—infraspinous fossa; U—vertebral border; A—axillary border; S—superior border; Ap—Acromion process; Sa—superior angle; Ia—inferior angle. Use this same code to label Figures 8-4, 8-5, and 8-6.

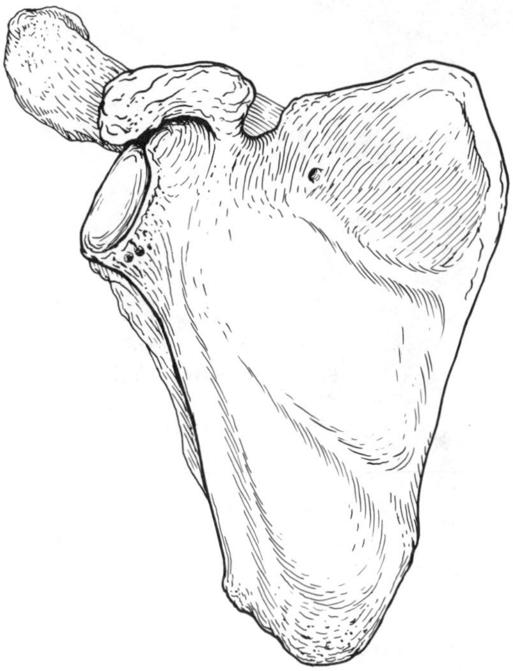

Figure 8-4. Right scapula from in front.

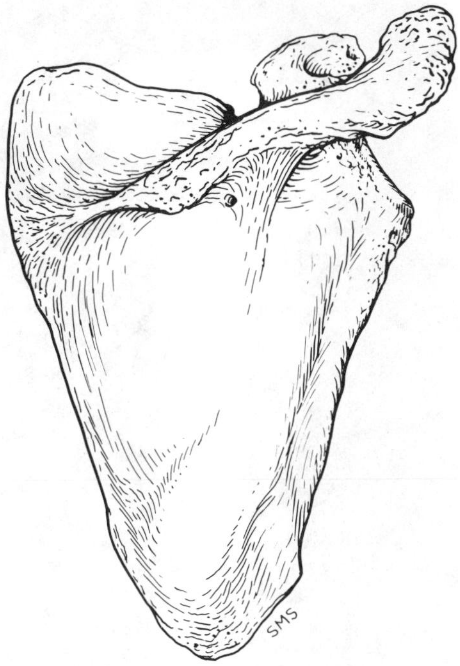

Figure 8-5. Right scapula from behind.

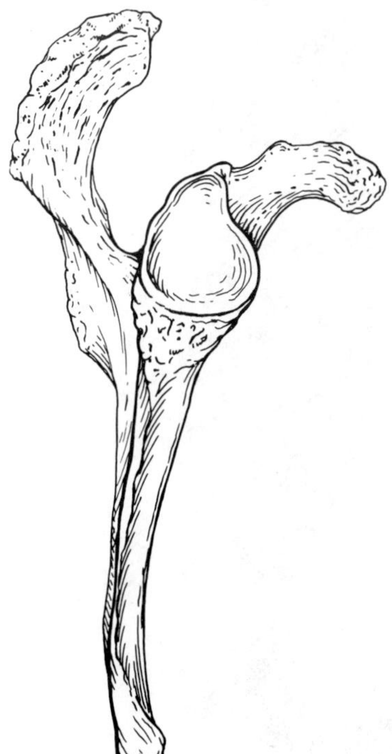

Figure 8-6. Three heavy bars of bone radiate from the glenoid fossa or cavity to three areas of the scapula where important muscles and ligaments are attached. (Right scapula, antero-lateral view.)

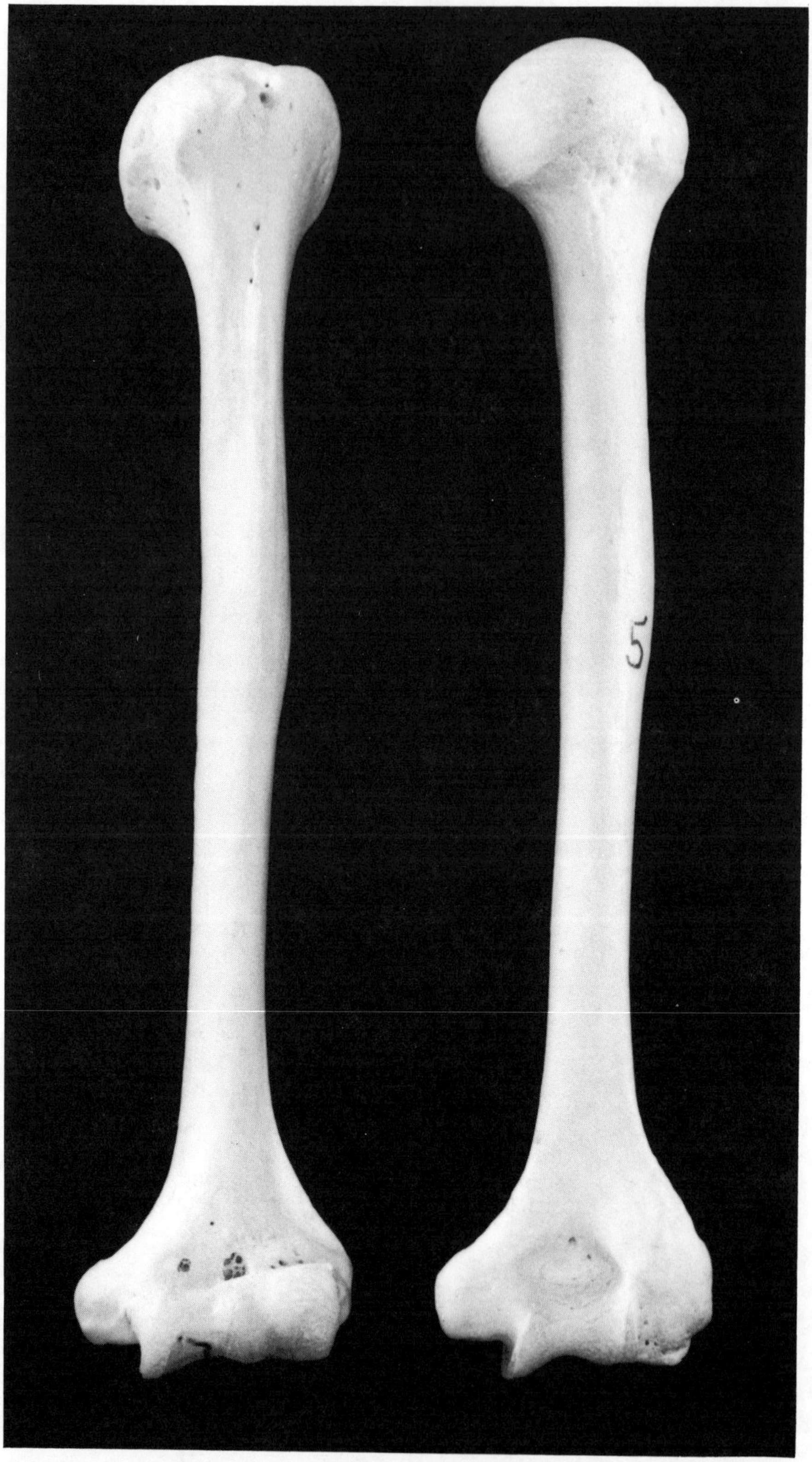

Figure 8-7. Left: ______ aspect of the ______ humerus. Right: ______ aspect of the ______ humerus. Indicate the following structures: H—head; B—bicipital groove; D—deltoid tubercle; M—medial epicondyle; L—lateral epicondyle; Tg—greater tuberosity; Tl—lesser tuberosity; C—coranoid fossa; O—olecranon fossa; Cp—capitulum.

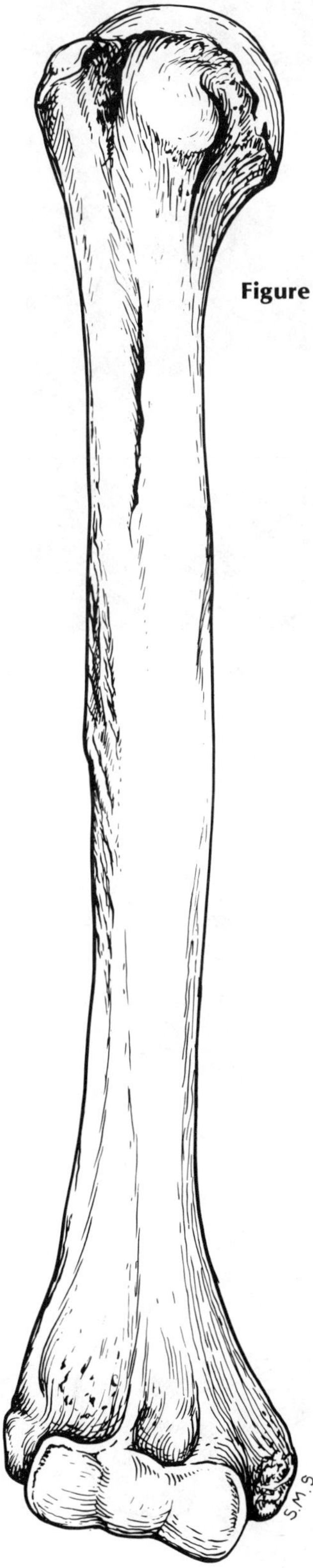

Figure 8-8. Right humerus from in front.

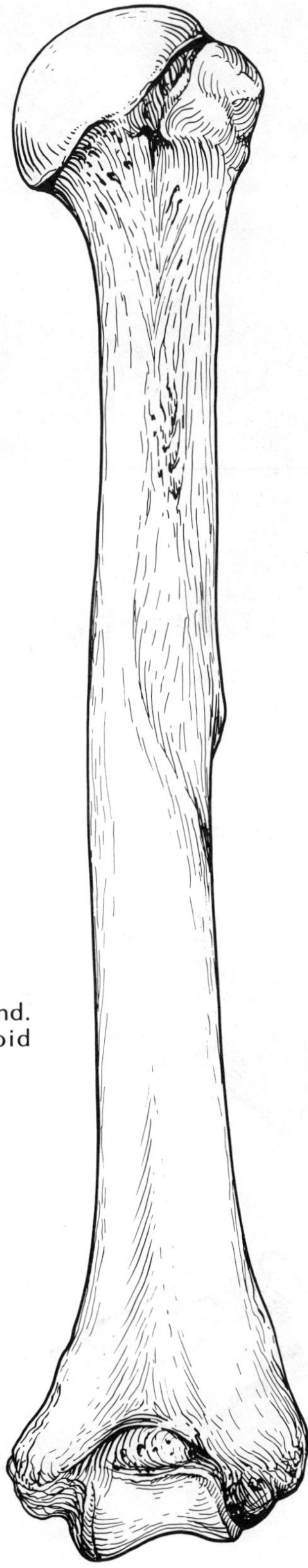

Figure 8-9. Right humerus from behind. Note spiral groove behind deltoid tubercle.

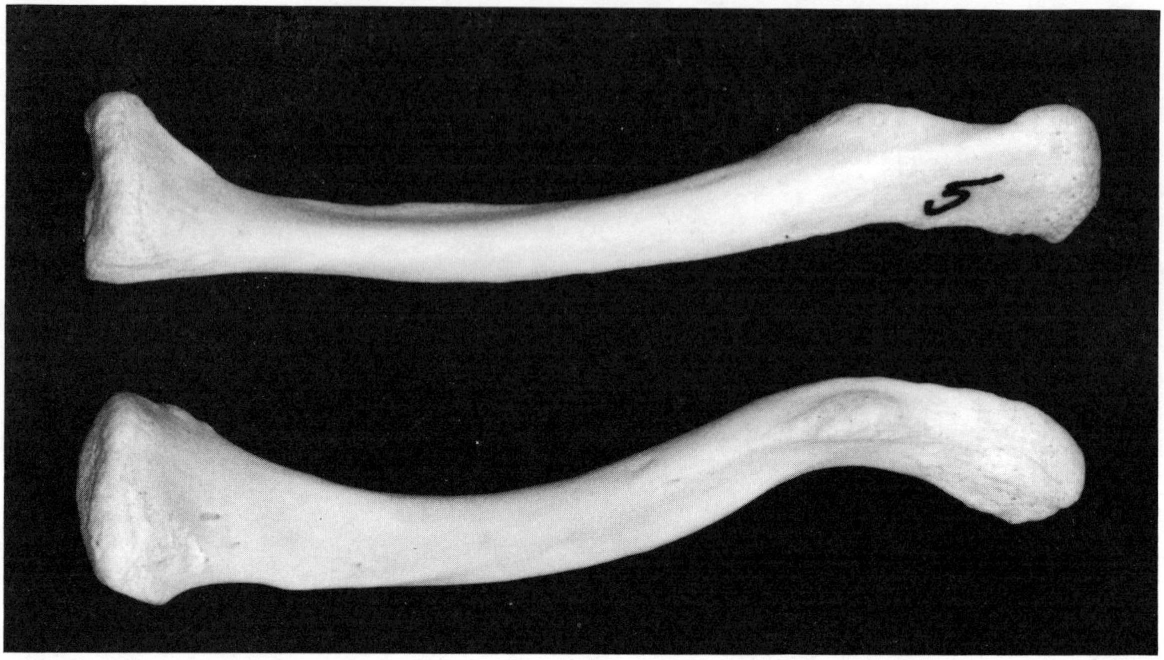

Figure 8-10. Top: right clavicle, superior aspect. Bottom: left clavicle, inferior aspect.

Figure 8-11. The clavicle. Top: lower surface. Bottom: upper surface.

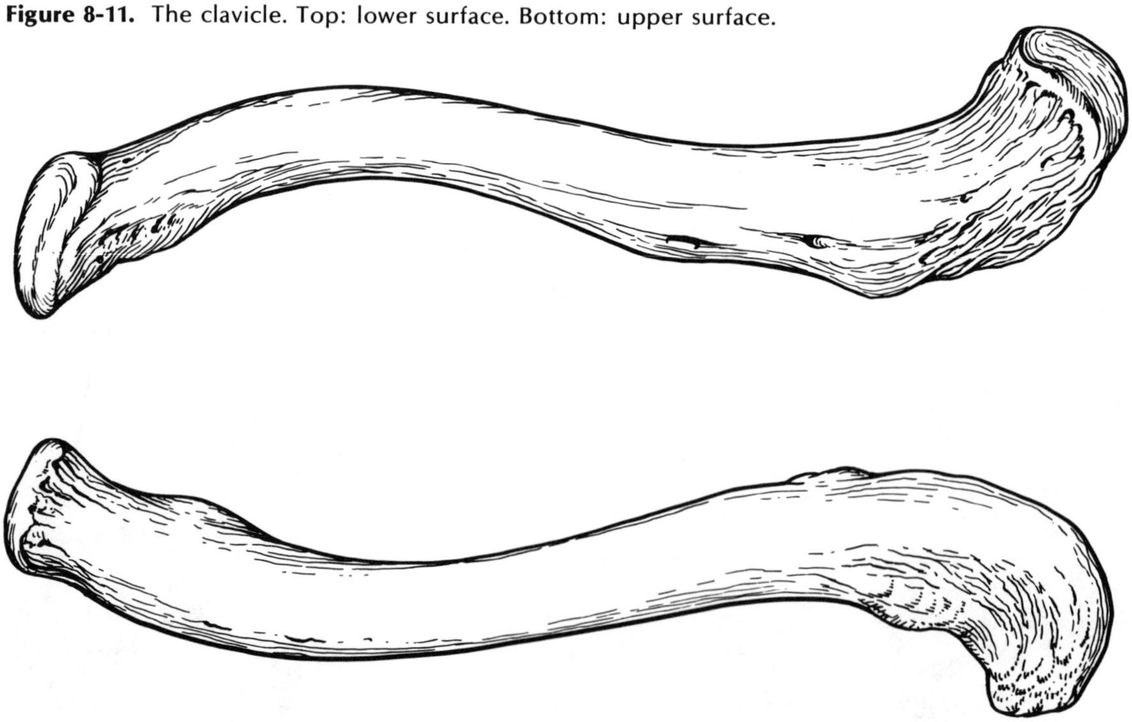

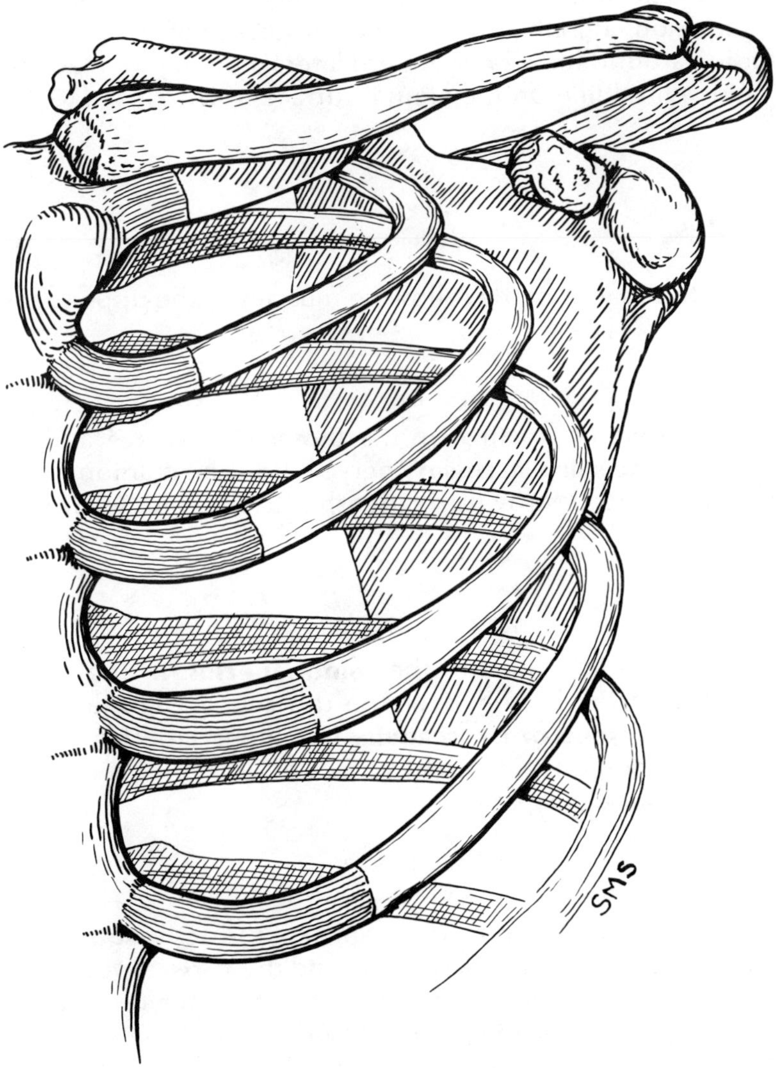

Figure 8-12. Orientations of bones of shoulder region. Use your text to aid sketching in the strong coraco clavicular ligament that suspends scapula from clavicle.

EXERCISE 1
DISSECTION

Make an incision along the mid-dorsal line from the base of the skull to the middle of the back. Extend the caudal portion of the incision laterally to the mid-ventral line. Extend the cranial end of the incision laterally down the dorsal surface of the forelimb to the elbow. Cut a circular incision around the elbow. A third lateral cut may be made from the back of the neck to the mid-ventral line on the ventral surface of the neck.

Superficial Muscles

Clear and locate the boundaries of the following muscles of the upper back and shoulder.

1. **Spinotrapezius**

   A flat, triangular muscle originating on the **thoracic vertebrae** and inserting on the **spine of the scapula**. The **spinal accessory** nerve enters the craniomedial surface of this muscle. Be careful not to destroy its relationship to the muscle. Transect this muscle along its origin.

2. **Acromiotrapezius**

   A flat muscle immediately craniad to the **spinotrapezius**. It originates from a heavy sheet of aponeurosis lying loosely over the thoracic vertebrae and inserts on the **acromion process** and **spine of the scapula**. It is innervated by the **spinal accessory nerve**, which can be seen on its internal surface after transection of the aponeurosis and reflexion of the muscle laterally.

3. **Clavotrapezius**

   This is a flat but somewhat thicker muscle craniad of the **acromiotrapezius**. It originates on the **lambdoidal ridge** of the **occipital bone** and the **cervical vertebrae**. Its fibers run laterally and curve forward over the front of the shoulder to insert on the vestigial **clavicle**. Innervation is by the **spinal accessory nerve** and the ventral rami of **C1-C4**. Careful dissection will expose this innervation.

4. **Clavodeltoid (Clavobrachialis)**

   A continuation of the **clavotrapezius** that originates on the **clavicle** and inserts on the medial surface of the **ulna**. Innervation is a branch of the **axillary nerve** that emerges from under the border of the **acromiodeltoid nerve** just caudal to and deeper than the **clavodeltoid**.

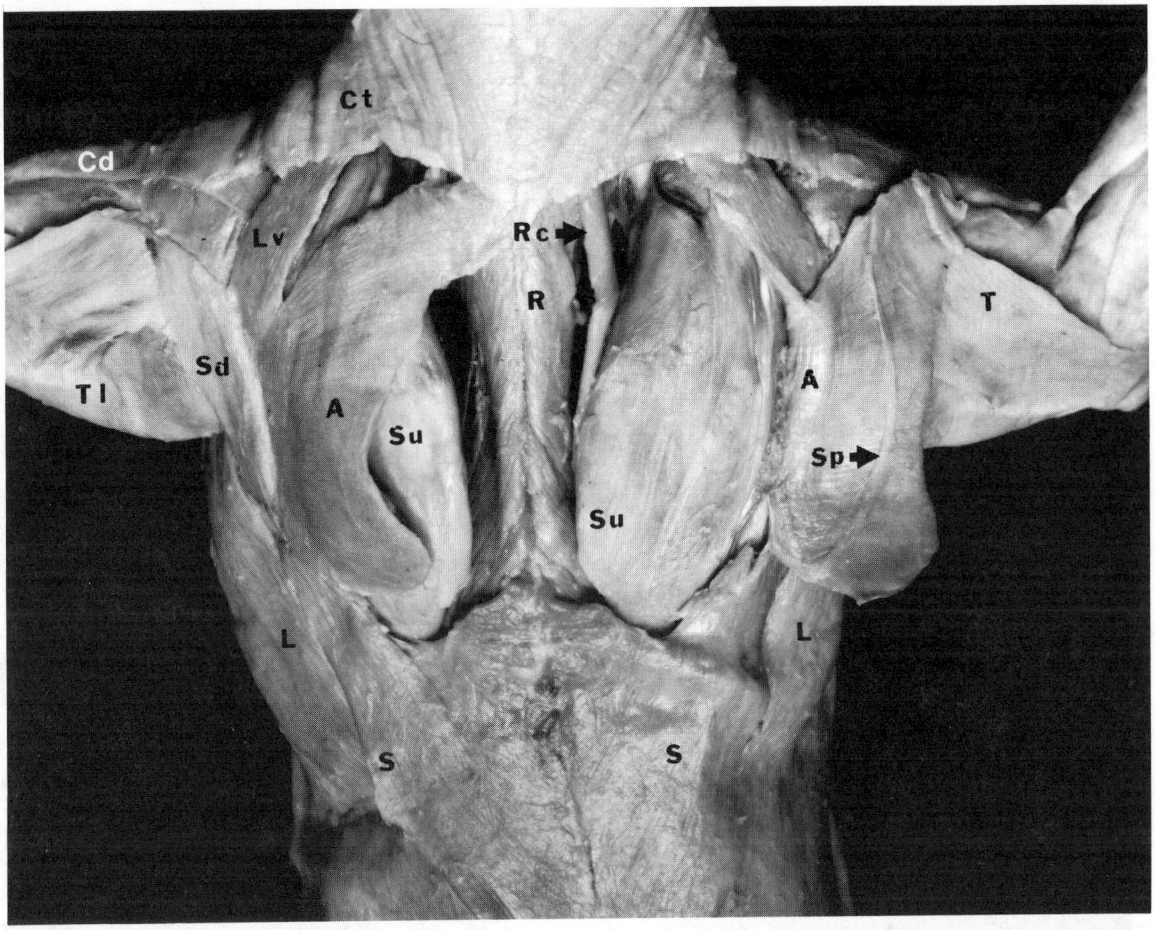

Figure 8-13. Superficial dissection of the back. S—spinotrapezius; Ct—clavotrapezius; Cd—clavodeltoideus; Tl—long head of triceps; Lv—levator scapulae ventralis; R—rhomboideus; Rc—rhomboideus capitus; Sp—spinal accessory; A—acromiotrapezius; Sd—spinodeltoideus; T—lateral head of triceps; L—latissimus dorsi; Su—supraspinatus.

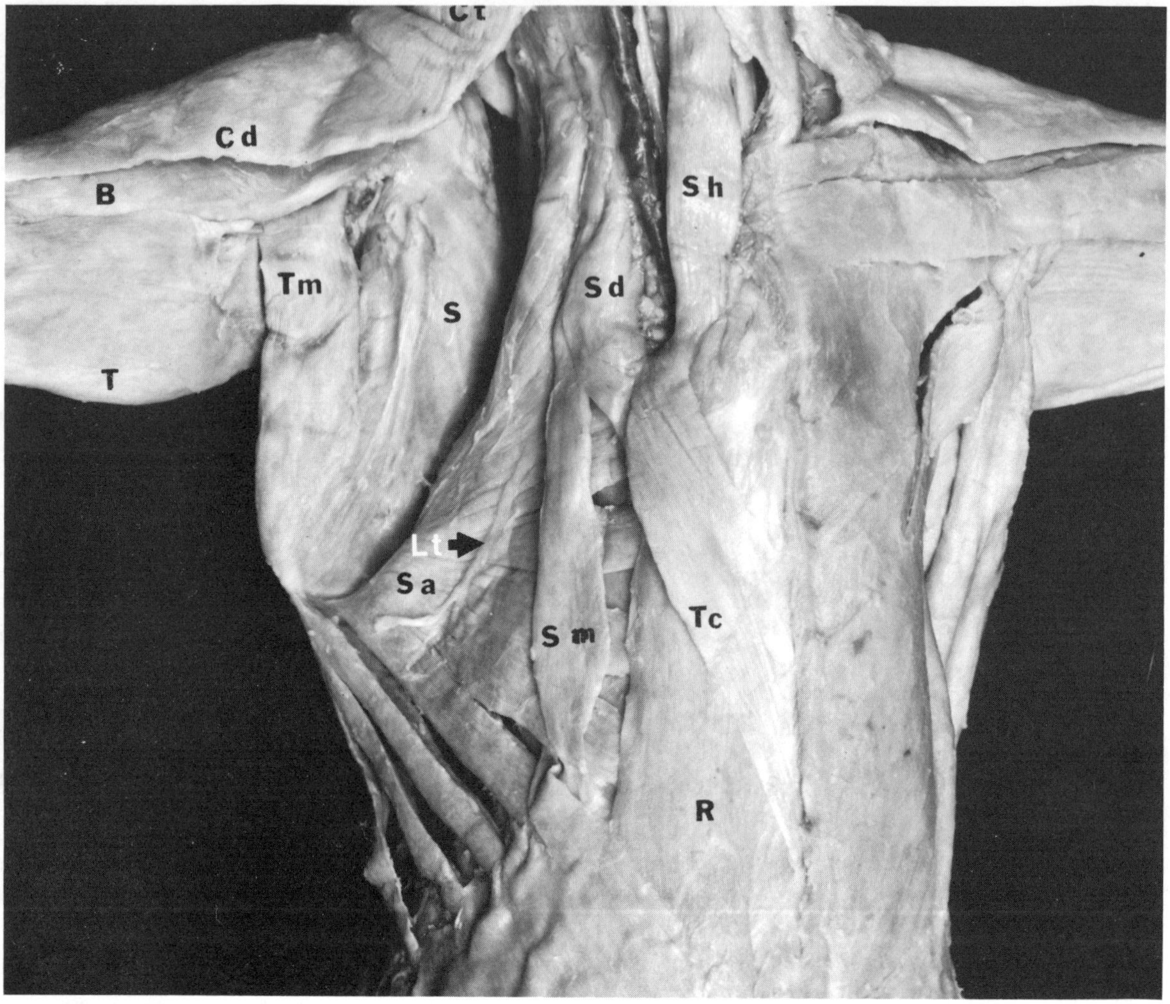

Figure 8-14. A deep dissection of the back. Ct—clavotrapezius; B—biceps femoris; Tm—teres major; Lt—long thoracic nerve; S—supraspinatus; Sa—serratus anterior; Sm—scalene medialis; Sd—scalene dorsalis; Sh—sternohyoid; Tc—transversus costarum; Cd—clavodeltoideus; T—long head of triceps; R—rectus abdominus.

5. **Acromiodeltoid**

   A thick, triangular muscle lying caudal to and underneath the **clavodeltoid**. It originates on the **metacromion** and **acromion process** of the **scapula** and inserts on the **deltoid** ridge of the **humerus**. Innervation—**axillary nerve**.

6. **Spinodeltoid**

   A thick muscle lying just caudal to the **acromiodeltoid** and partially underneath the fibers of it. Its ventral fibers overlie the massive, rounded bulk of the long head of the **triceps br**. Origin—**spine of the scapula**. Insertion—**deltoid ridge** of the **humerus**, just dorsal to the insertion of the **acromiodeltoid**. Innervation—branch of the **axillary nerve**.

7. **Levator scapulae ventralis**

   A long slender muscle with fibers running caudad and ventrad from its origin on the wings of the **atlas** and the **occipital** bone to its insertion on the **metacromion process** of the **scapula**. The dorsal border of the **acromiodeltoid** and the ventral border of the **acromiotrapezius** are adjacent to this muscle on its ventral and dorsal borders. It is innervated by **C-3** and **C-4**.

8. **Latissimus dorsi**

   This large, sheetlike muscle covers the flank and sweeps forward, twisting ventrally to insert on the shaft of the **humerus** with its fibers inseparably fused with those of the muscles of the thorax. It is innervated by the **anterior thoracic** and **subscapular nerves**.

Deep Muscles

Transect the **latissimus dorsi**, the **spinotrapezius**, and the **acromiotrapezius**. Hold the forelimb of the cat in complete adduction in order to stretch the **rhomboids** between the vertebral border of the **scapula** and the **vertebral** column. Locate and dissect the following muscles.

1. **Rhomboideus**

   A thick, rhomboid-shaped muscle originating on the **cervical thoracic vertebrae** and inserting along the vertebral border of the **scapula**. It is innervated by ventral rami of thoracic spinal nerves.

2. **Rhomboideus capitus**

   A thin, elongated muscle attached to the cranial border of the **rhomboideus**.

3. **Supraspinatus**

   A thick muscle occupying the entire **supraspinous fossa**, from which it originates. It inserts with other rotator cuff muscles on the **humerus**, specifically on the greater tuberosity. Innervation is by the **suprascapular nerve**.

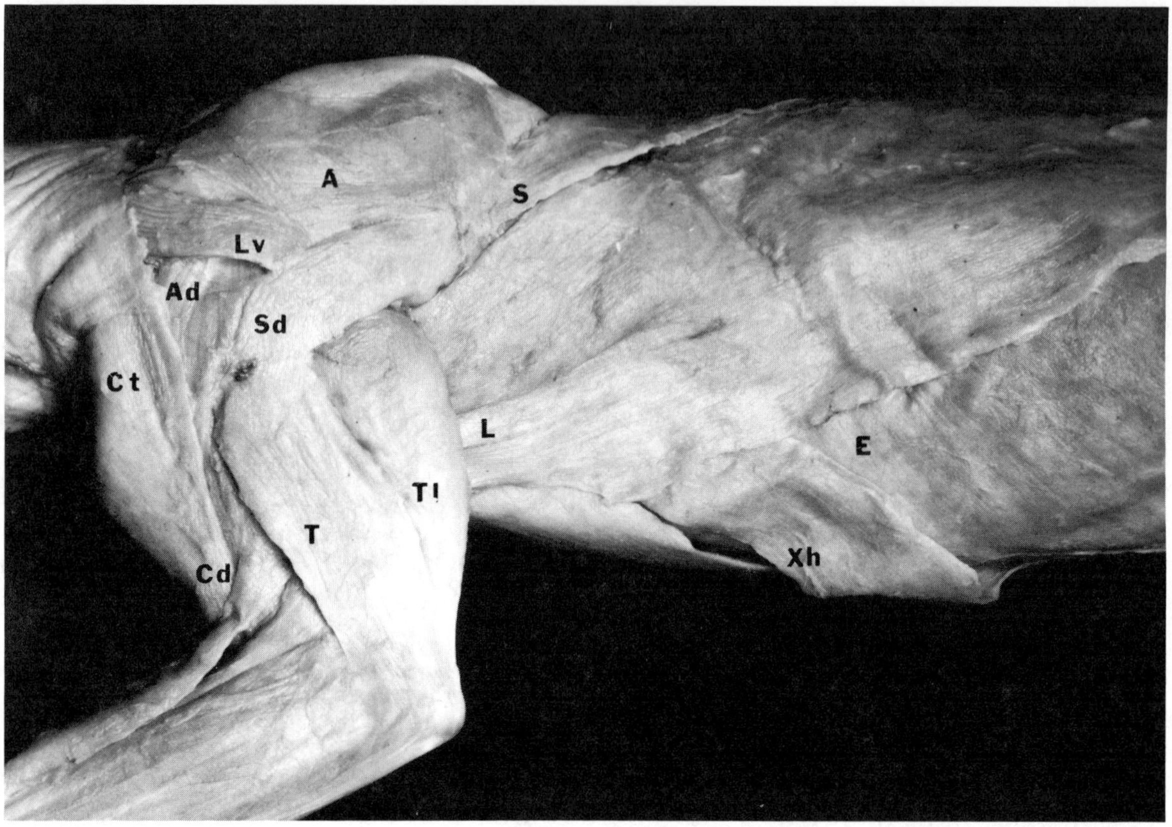

Figure 8-15. Lateral view of shoulder region. A—acromiotrapezius; Ct—clavotrapezius; Sd—spinodeltoideus; S—spinotrapezius; Cd—clavodeltoideus; S—spinotrapezius.

4. **Infraspinatus**

   In the cat, this muscle is smaller than the **supraspinatus**. It originates in the **infraspinous fossa** and inserts on the **greater tuberosity** of the **humerus**. It is also innervated by the **suprascapular**.

5. **Teres minor**

   This is a small muscle running along between the glenoid border of the **scapula** and the **humerus**. It consists of a small, fleshy portion situated between two thick tendons. It is innervated by the **axillary nerve**.

6. **Subscapularis**

   This triangular muscle occupies most of the **subscapular fossa**, from which it originates. It inserts on the lesser tuberosity of the **humerus**. Innervation is by the **subscapular nerve**.

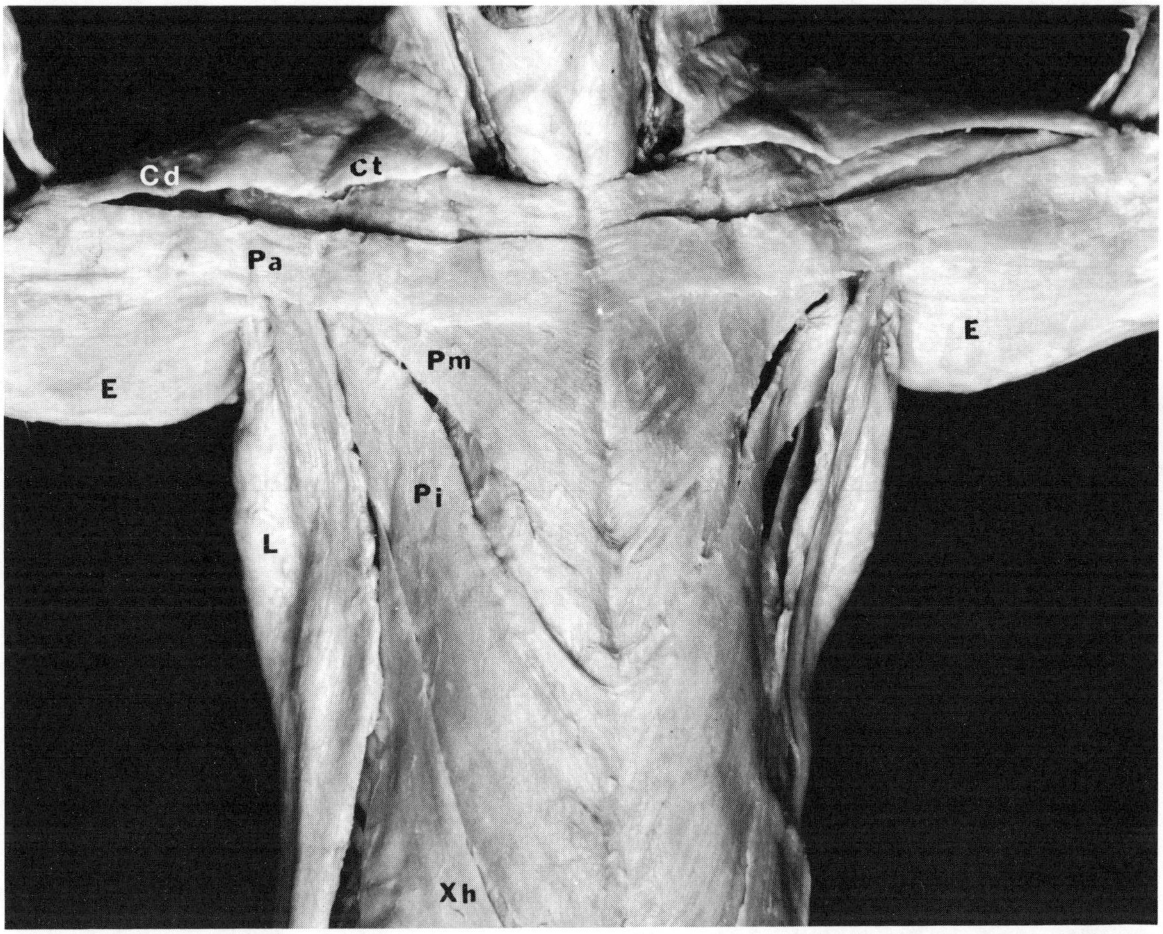

Figure 8-16. Superficial dissection of the ventral thorax. Cd—clavodeltoideus; Pa—pectoantebrachialis; Pi—pectoralis minor; Xh—xiphihumeralis; Ct—clavotrapezius; Pm—pectoralis major; L—latissimus dorsi; E—epitrochlearis.

7. **Teres major**

   In the cat, this is a large and important muscle. It is strategically situated to aid in forceful extension of the **humerus**, an important action in running. It originates from the **glenoid process** on the medial border of the **scapula** and inserts on the pectoral ridge of the surface of the **humerus**. It is innervated by the **subscapular nerve.**

8. **Serratus anterior (Serratus ventralis)**

   A fan-shaped muscle originating on the first nine or ten ribs and inserting on the vertebral border of the costal surface of the **scapula**. It furnishes a slinglike arrangement with the two front legs—an arrangement that holds the trunk of the cat in position. Innervation is by the **posterior thoracic nerve**.

Pectoral Muscles

Clear and locate the boundaries of the following muscles of the thorax and upper arm.

1. **Pectoantebrachialis**

   A thin, flat, straplike muscle overlying the cranial border of the **pectoralis major**. Determine its origin and insertion. It is innervated by the **anterior thoracic nerve**.

2. **Pectoralis major**

   A fairly thick, triangular muscle originating on the **manubrium** and **sternum** and inserting on the pectoral ridge of the **humerus**. Innervation—**anterior thoracic nerve**.

3. **Pectoralis minor**

   A thin, straplike muscle with its fibers running forward and laterally from its origin on the caudal portion of the **sternum** and **xiphoid process**. It inserts on the pectoral ridge of the **humerus**. Innervation—**anterior thoracic nerve**.

4. **Xiphihumeralis**

   A thin band of muscle lying just caudal to the **pectoralis minor**. It originates on the **xiphoid process** and inserts on the **humerus** by a fusing of fibers with both the **latissimus dorsi** and **pectoralis minor**. Innervation—**anterior thoracic nerve**.

Exposing the Axilla

1. Transect the following muscles being sure that no major nerves or blood vessels are cut by mistake.

   Pectoantebrachialis
   Xiphihumeralis
   Pectoralis minor
   Pectoralis major
   Clavodeltoideus

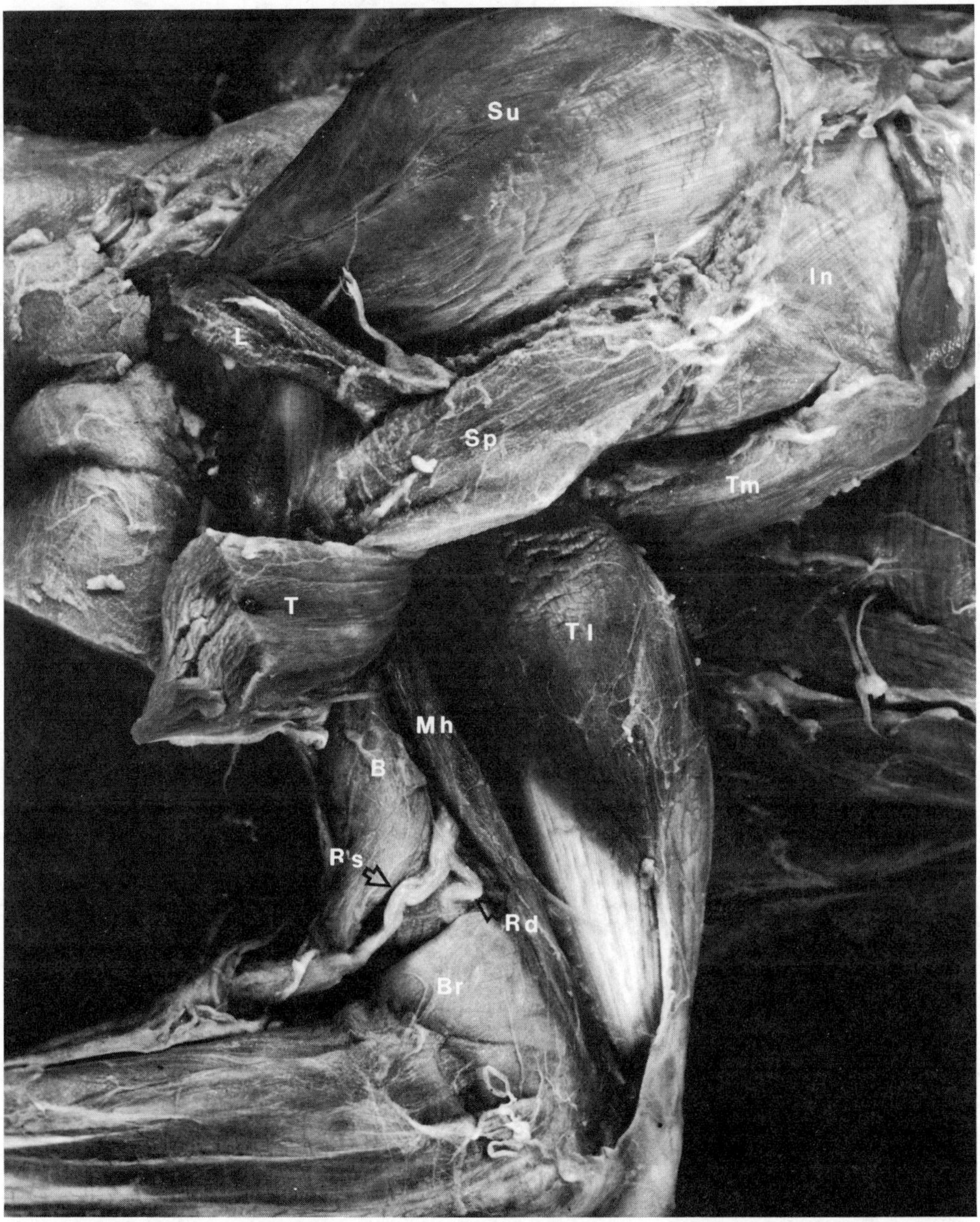

Figure 8-17. Dissection of shoulder region, lateral view. Muscles: Su—supraspinatus; In—infraspinatus; L—levator scapuliventralis; Sp—spinodeltoideus; Tm—teres major; T—lateral head of triceps; Tl—long head of triceps; B—brachialis; Rs—superficial branch of radial nerve; Rd—deep branch of radial nerve; Br—brachialradialis.

2. Reflect the insertions back so that the **biceps brachii** and **epitrochlearis** are exposed.

3. Transect the **biceps brachii**, but be careful not to cut the **musculocutaneous nerve** entering the proximal end of the muscle from the posterior medial side.

4. Transect the **epitrochlearis**, again being careful not to cut the small nerve that enters it (a branch of the **radial nerve**).

5. The **axilla** is now exposed, and the sheath of the **brachial plexus**, the **axillary**, and **brachial** arteries and veins is visible. The primary structures are heavily wrapped in loose connective tissue with large fat deposits, which must be cleared away before the individual nerves can be seen. The **musculocutaneous nerve** can be isolated by finding the branch that enters the **biceps brachii** and tracing it back to the main nerve. Likewise, the **radial nerve** can be identified by locating the small branch that innervates the **epitrochlearis** and tracing it back to the source.

   Do not attempt to locate the **ulnar** and **median nerves** at this time. They will be more easily identified later.

6. **Coracobrachialis**

   This is a small, spindle-shaped muscle originating on the **coracoid process** and inserting on the **medial surface** of the proximal end of the **humerus**. Ascertain its innervation.

7. Locate the three heads of the **triceps**. When they are transected, use care to avoid cutting the **radial nerve** which lies between the lateral and intermediate parts.

   The **anconeus** forms a fourth head of the **triceps** in the cat and can be observed after transecting the **lateral head**.

8. Find the **brachialis** under the **biceps** and inserting on the **ulna**. What is its innervation?

EXERCISE 2
DISCUSSION QUESTIONS

1. How do the following items in the cat differ from those in man? What is the significance of this difference?

   a. Position of scapula on rib cage:

   ____

   ____

b. Relative length and width of scapula.

____

____

c. Size and attachments of clavicle.

____

2. How do the triceps of the cat compare to those of man? Explain this difference in view of the function of these muscles in the two species.

____

____

LABORATORY 9

Dissection of Cat Forelimb II: Forearm

OBJECTIVES

By the end of this laboratory period, you should be able to identify the following.

Bones

Ulna
Radius
Carpals
Metacarpals

Nerves

Median
Radial
Ulnar

Muscles

Extensor Area
Brachioradialis
Extensor carpi radialis longus
Extensor carpi radialis brevis
Extensor digitorum communis
Extensor digitorum lateralis
Extensor carpi ulnaris
Extensor indicus

Flexor Area
Palmaris longus
Flexor carpi ulnaris
Flexor carpi radialis
Pronator teres
Flexor digitorum profundus
Pronator quadratus
Flexor digitorum sublimus

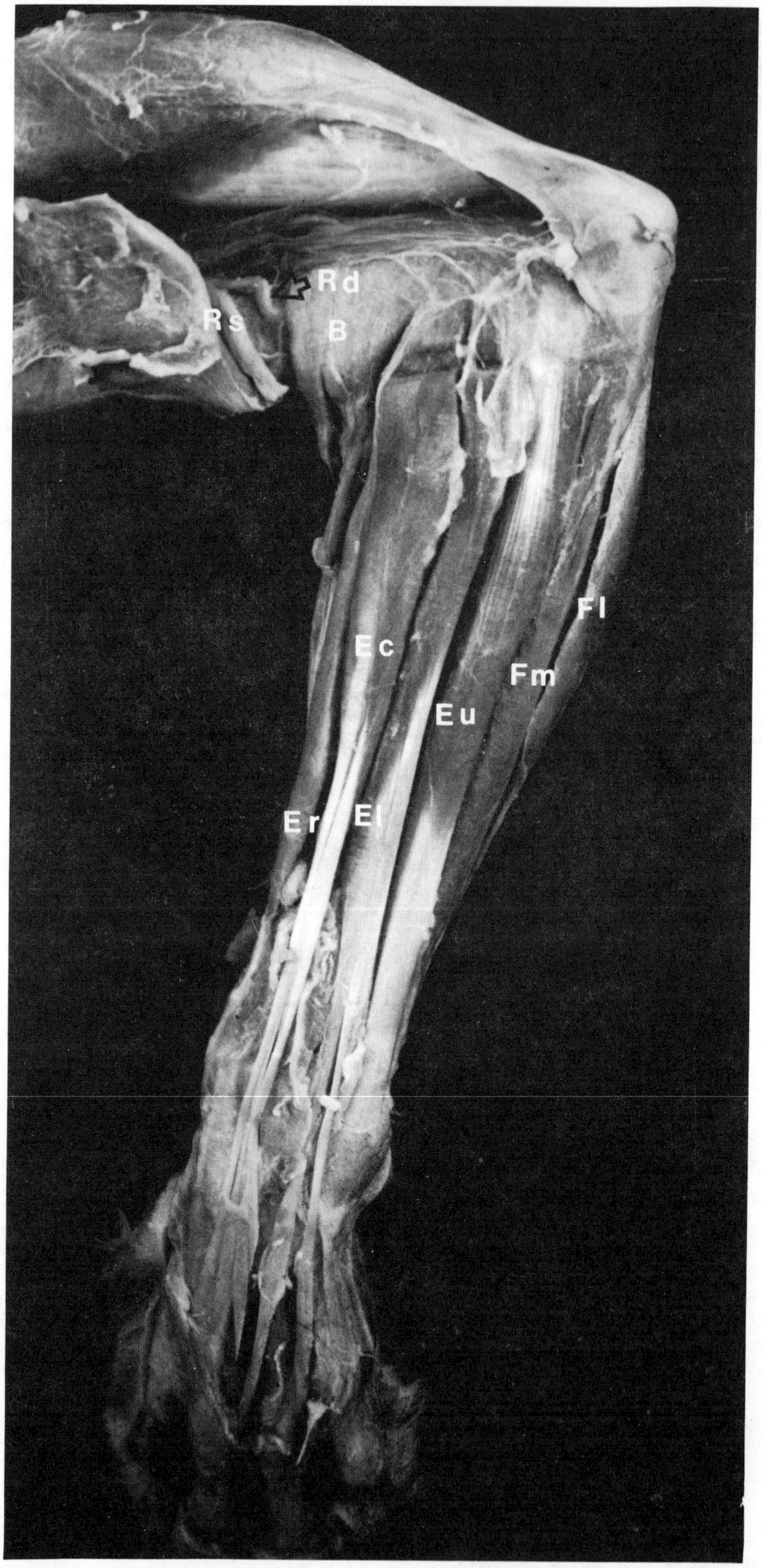

Figure 9-1. Dissection of cat forearm, extensor region. Muscles: B—brachioradialis; Er—extensor carpi radialis; Ec—extensor digitorum communis; El—extensor digitorum lateralis; Eu—extensor carpi ulnaris; Fm—flexor carpi ulnaris; medial head; Fl—flexor carpi ulnaris, lateral head. Nerves: Ra—radial nerve and superficial branch (cut); Rd—radial nerve, deep branch.

EXERCISE 1
DISSECTION OF POSTERIOR SURFACE—EXTENSOR AREA

1. Remove the skin from the entire forearm and paw.

2. Observe the tough coating of connective tissue covering the forearm. Insert a scalpel between any two muscles on the posterior surface of the forearm, and carefully split and pull off this connective tissue.

3. The key to the extensor region is the **brachioradialis**, which passes along the lateral surface of the forearm from its origin on the posterior surface of the middle of the shaft of the **humerus**. This long, thin muscle is wrapped in a connective tissue sheath which also contains the superficial branch of the radial nerve.

   After finding the **brachioradialis**, locate the following superficial extensors by carefully dissecting out their tendons of insertion. They are listed in order from medial to lateral.

   a. **Extensor carpi radialis—longus and brevis**
   Find the deep **radial nerve** passing between these two muscles.

   b. **Extensor digitorum communis**

   c. **Extensor digitorum lateralis**

   d. **Extensor carpi ulnaris**
   Find the origin, insertion, and innervation of these muscles.

4. **Deep Extensors**

   Locate the following two muscles from the description below and from the detailed demonstration dissections.

   a. **Extensor indicus** is a long, flat muscle under the **extensor carpi ulnaris**. The distinguishing feature is a long, flat tendon that runs parallel to the **radius** nearly to the wrist, then passes obliquely under the **extensor carpi ulnaris** and **extensor digitorum lateralis** to insert on the intermediate phalanx of the index digit.

   b. The **extensor pollicus brevis** muscle extends from the lateral surface of the **ulna** to the first **metacarpal**. The fibers run obliquely down and mediad to the broad, flat tendon of insertion.

   What nerve supplies these muscles?

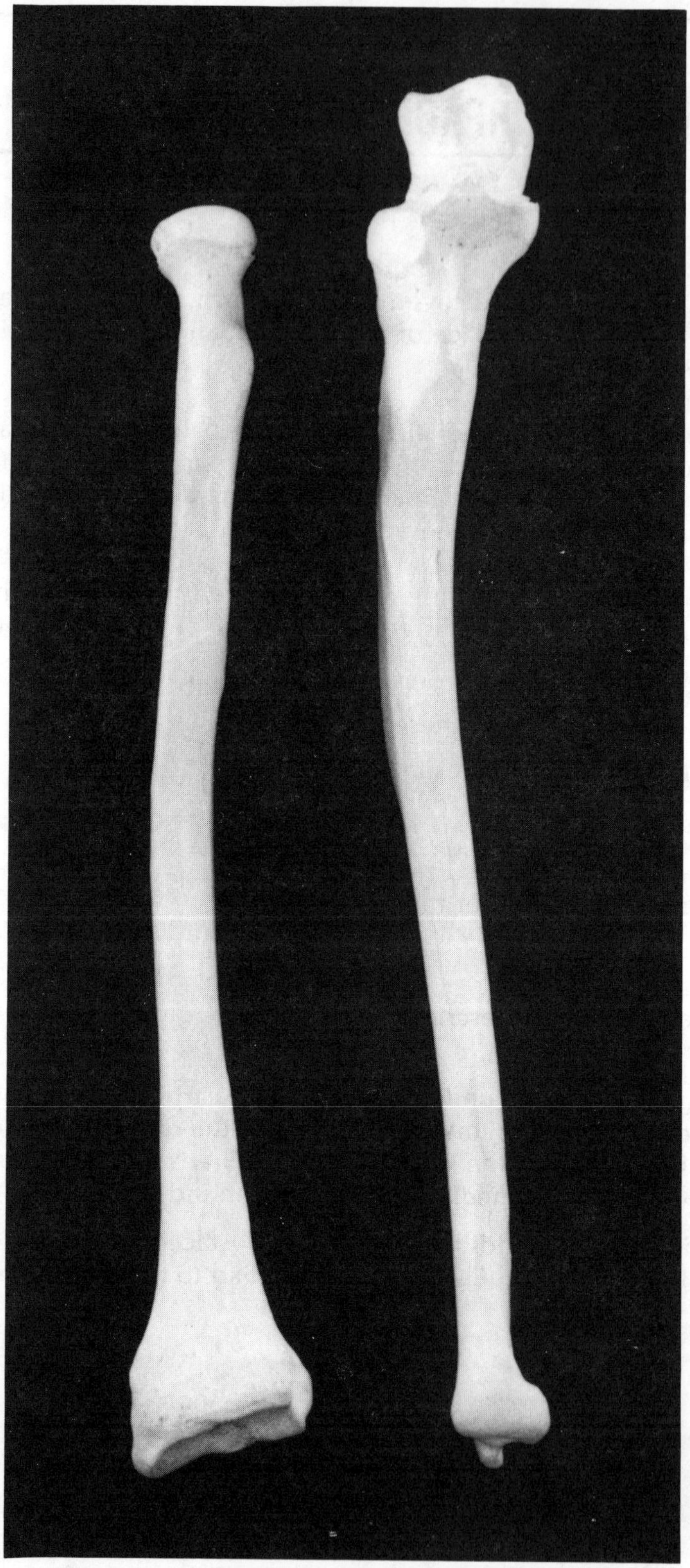

Figure 9-2. Label the structures according to the following code. O—olecranon process; C—coranoid process; T—trochlear notch; R—radial notch; Cr—crest; S—styloid process; H—head; B—bicipital tubercle; C—crest; S—styloid process. Use this same code to label Figures 9-3 and 9-4.

Figure 9-3. Right radius and ulna, anterior (front) view.

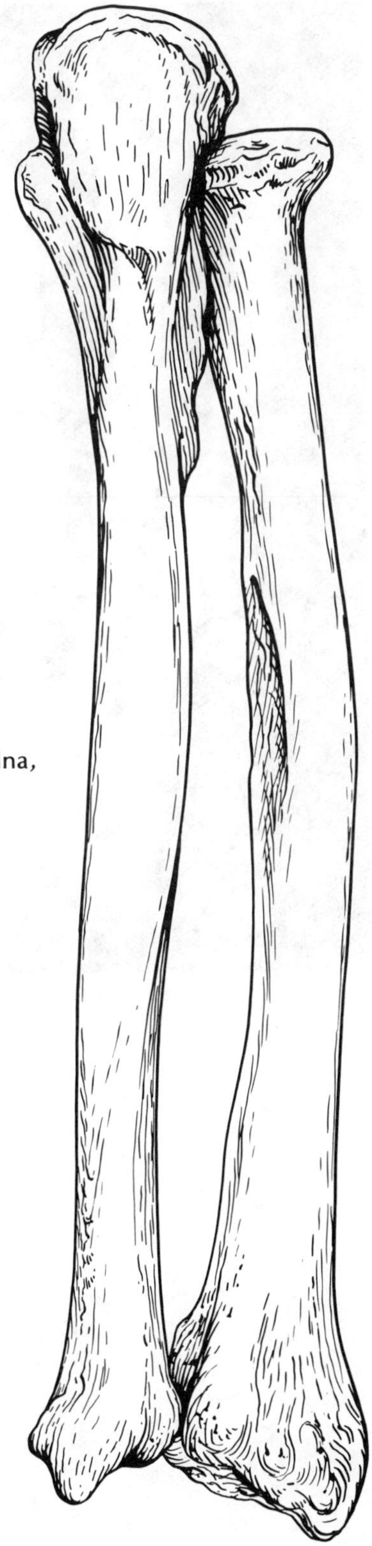

Figure 9-4. Right radius and ulna, posterior (back) view.

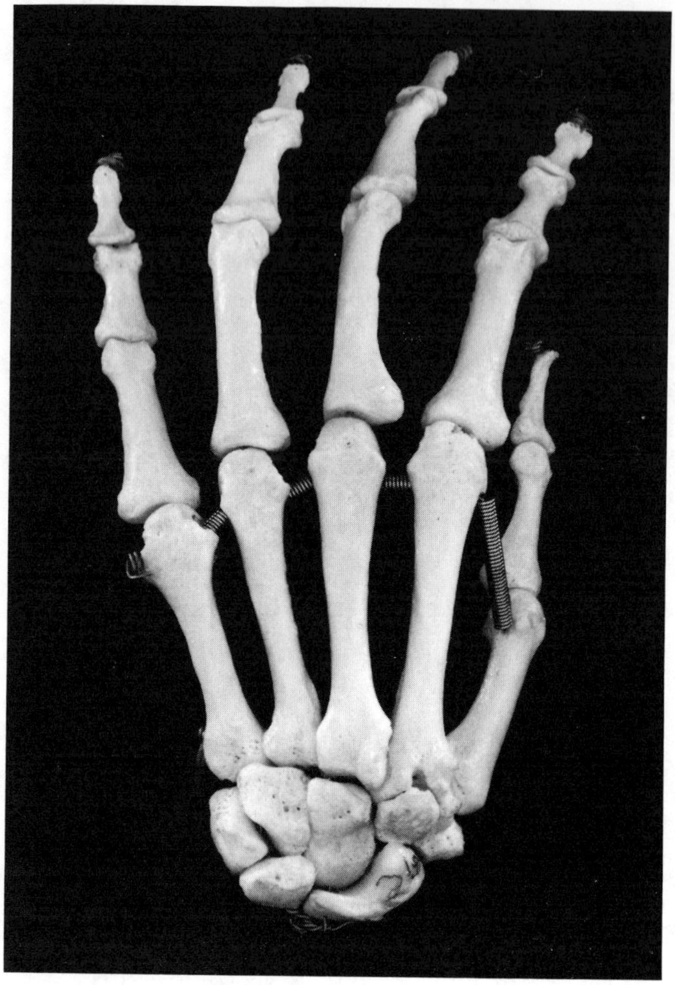

Figure 9-5. Dorsal aspect of the left hand. Label the indicated structures according to the suggested code. P—pisiform; Tr—triquetrum; L—lunate; S—scaphoid; H—hamage; C—capitate; Tp—trapezoid; Tz—trapezium; Ml—metacarpals; Pp—proximal phalanges; Pm—middle phalanges; Pd—distal phalanges.

Figure 9-6. Dorsal view.

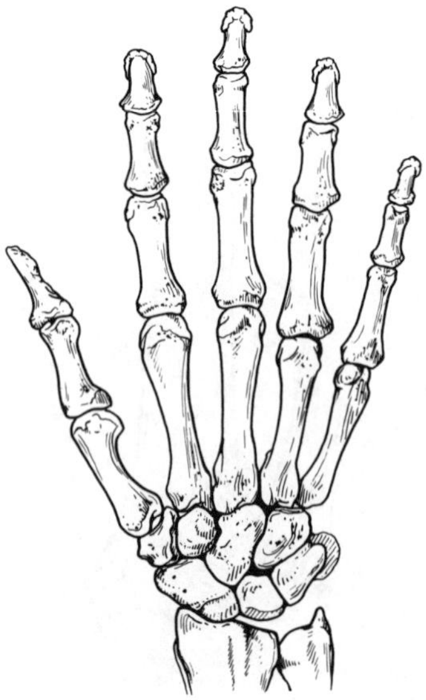

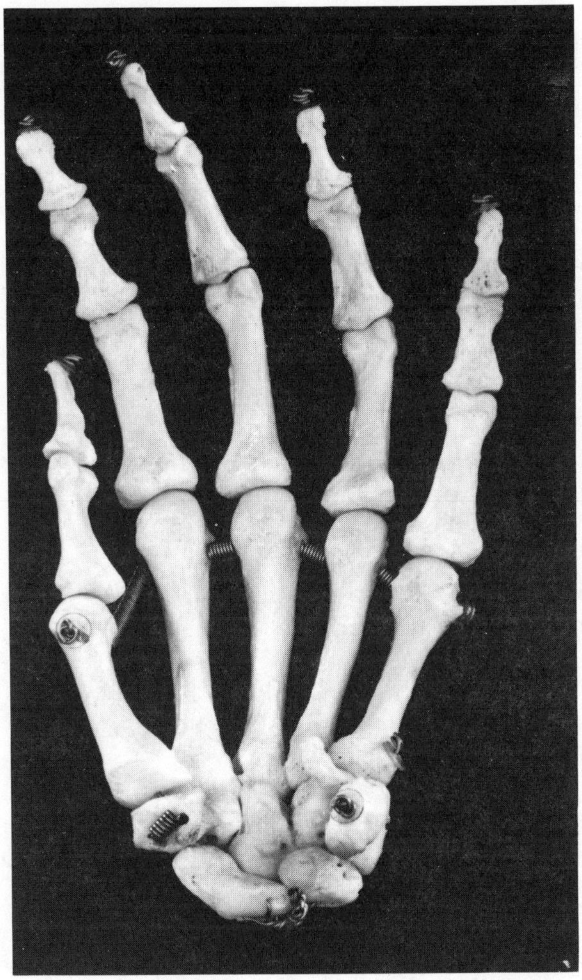

Figure 9-7. ________ ventral aspect of the left hand. Label the indicated structures according to the suggested code. P—pisiform; Tr—triquetrum; L—lunate; S—scaphoid; H—hamage; C—capitate; Tp—trapezoid; Tz—trapezium; MI—5 metacarpals; Pp—proximal phalanges; Pm—middle phalanges; Pd—distal phalanges.

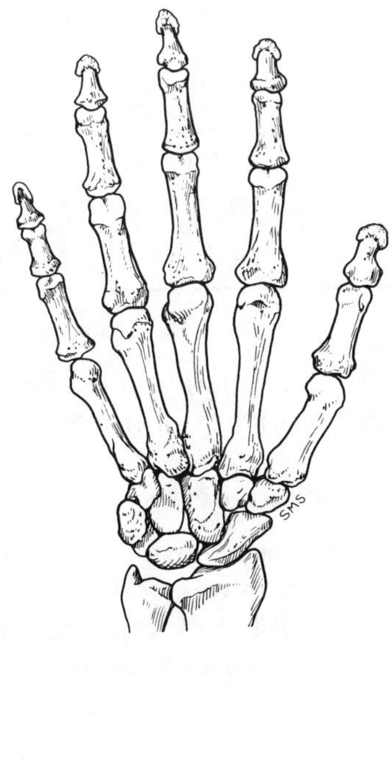

Figure 9-8. Ventral (palmar) view.

EXERCISE 2
DISSECTION OF THE FLEXOR REGION

1. The key to the flexor area is the broad, flat **palmaris longus**, which occupies most of the center of the length of the forearm on the anterior aspect. This muscle originates on the **medial epicondyle** and splits into four tendons of insertion.

2. The **flexor carpi ulnaris** possesses two heads, one arising from the **medial epicondyle** of the humerus, the other from the medial side of the **olecranon process**.

3. The **flexor carpi radialis** lies just medial to the **palmaris longus**. It originates on the **medial epicondyle** and inserts on the distal end of the **radius**. The **radial artery** is attached to the lower half of this muscle.

4. **Pronator teres** is the muscle that lies medial to the **FCR** and originates from the same area. It narrows to a slim, flat tendon of insertion on the **radius**.

5. The **flexor digitorum profundus** has five heads inserting on a thick, heavy tendon which passes over the front of the wrist deeper than the tendons of the **FCR**, **FCU**, and the **PL**. With the aid of the instructor, identify these heads.

6. The **pronator quadratus** occupies the anterior surface of the distal ends of the **radius** and **ulna**. It is the deepest muscle on the distal end of the forearm and may be seen only after the muscles listed above have been transected.

7. If time permits, locate the **flexor digitorum sublimus**, which originates in two portions. One arises from the outer border of the tendon of the **palmaris longus**, and the other arises from the superficial aspect of the tendon of the FDP.

EXERCISE 3
DISSECTION OF THE NERVES

With the separation of the muscles of the arm, it should now be possible to locate all five main branches of the **brachial plexus**.

1. **Axillary nerve**

   This nerve extends ventro-lateral through the axilla and emerges laterally between the **teres minor** and the ventral border of the proximal end of the head of the **triceps brachii**. Branches run to the interior surface of the three **deltoid muscles**. These branches can be traced back to identify the nerve.

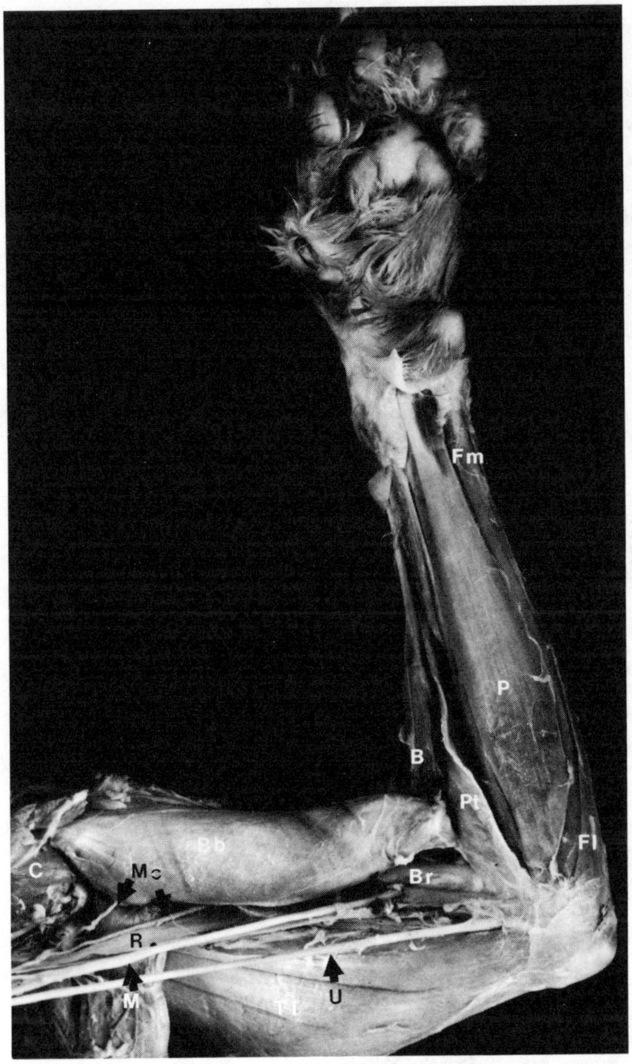

Figure 9-9. Dissection of cat forearm, superficial muscles of flexor region. Muscles: B—brachioradialis; P—palmaris longus; Fm—flexor carpi ulnaris, medial head; Fl—flexor carpi ulnaris, lateral head; Pt—pronator teres; Bb—biceps brachii; Tl—long head of triceps; Br—brachialis; C—coracobrachialis. Nerves: U—ulnar; M—median; Mc—musculocutaneous; R—radial.

2. **Musculocutaneous**

   This nerve is the most anterior of the large nerves of the **brachial plexus**. It gives branches off to the **biceps** and **coracobrachialis**, after which it passes across the forearm for skin distribution in the forearm.

3. The **Radial nerve**

   This nerve arises from the thick spinal nerves C-7 and 8 and T-1. A slender branch is given off to the **epitrochlearis** almost from the origin of the nerve. The nerve emerges from between the heads of the triceps and is easily observed when the lateral head is transected. The **radial nerve** splits above the **humerus** into the deep and superficial **radial nerves**. The former enters the forearm between the distal end of the **brachialis muscle** and the proximal end of the **extensor carpi radialis**. It passes under the **extensor digitorum communis** and over the **supinator**.

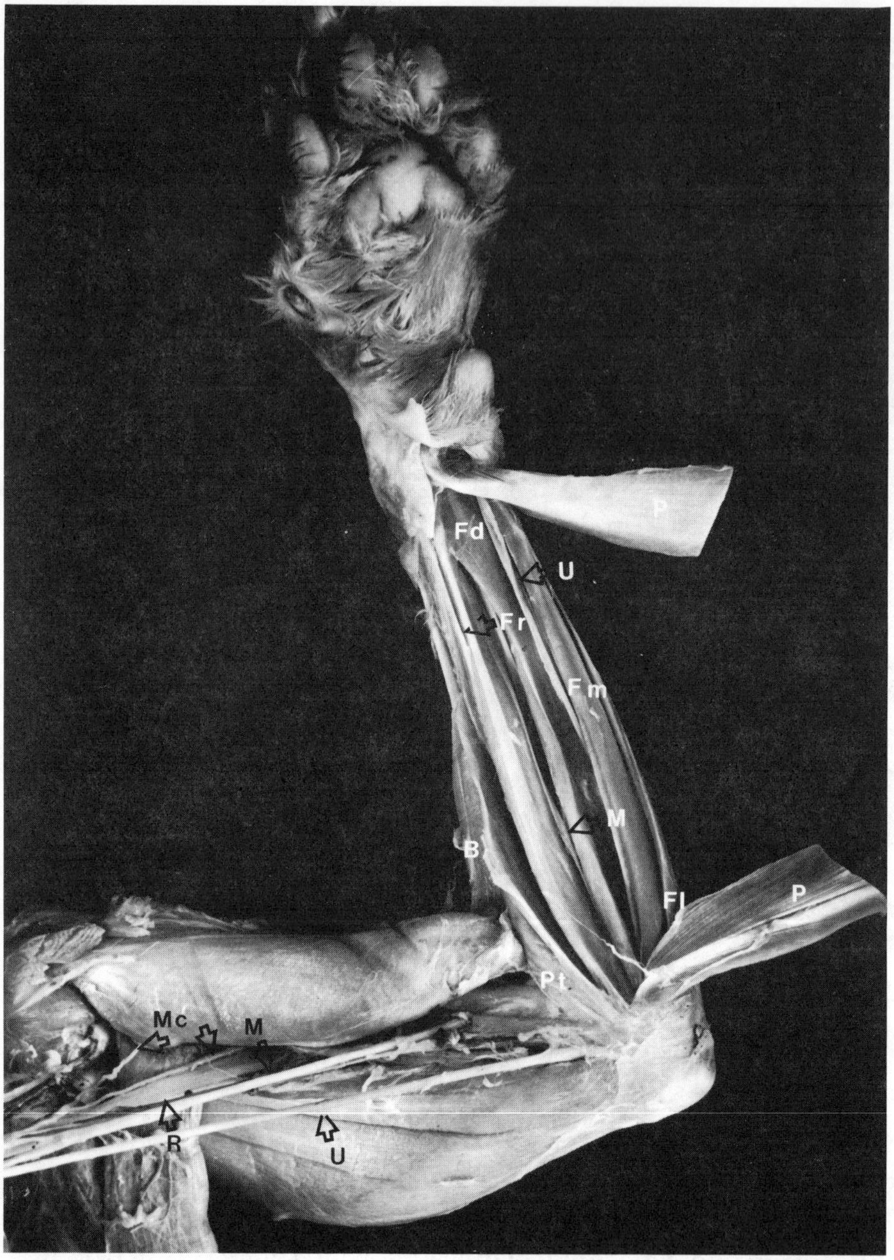

Figure 9-10. Dissection of cat forearm, palmaris longus transected. Muscles: B—brachioradialis; P—palmaris longus; Fm—flexor carpi ulnaris, medial head; Fl—flexor carpi ulnaris, lateral head; Pt—pronator teres; Fr—flexor carpi radialis tendon; Fd—flexor digitorum profundus tendon. Nerves: U—ulnar; M—medial; Mc—musculocutaneous; R—radial.

4. The **ulnar nerve** divides into two branches as it passes around the elbow at the **medial epicondyle**. The principal branch lies between the heads of the FCU.

5. The **medial nerve** enters the forearm through the **epicondylar foramen** and divides into two branches, one supplying the PT, FDP, and the PL, the other passing across the rest of the digits.

EXERCISE 4
DISCUSSION QUESTIONS

1. Which group of muscles—the extensor or the flexors of the wrist—seems to be the most highly developed in the cat? How can you explain this?

2. What muscle groups in the cat would be paralyzed by an injury to the radial nerve as it exits from the brachial plexus? The median? The ulnar?

3. How would the cat's gait be affected by the above situations?

Table 9-1

Classification of Joints

Joint Class	Variety	Nature of Union	Example
Immovable (Synarthroses)	1. Suture	Thin layer of fibrous tissue between bony surfaces.	Skull sutures
	2. Schindylesis	Bony plate inserted into bony cleft or fissure	Rostrum of Sphenoid and perpendicular plate of Ethmoid
	3. Gomphosis	Insertion of conical process into socket	Teeth in alveoli of Maxilla or Mandible
	4. Synchondrosis	Cartilage connecting two bony surfaces	First costosternal articulation
Slightly movable (amphiarthroses)	1. Symphysis	Bony surfaces connected by broad, flat disc of fibrocartilage	Intervertebral articulations
	2. Syndesmosis	Interosseous ligament	Inferior tibiofibular articulation
Freely movable (Diarthroses)	1. Hinge-joint (Ginglymus)	Joint cavity. Bony surfaces permit motion in only one plane	Humeroulnar articulation
	2. Pivot-joint (Trochoid)	Joint cavity. Bony, pivotlike process fits into receiving ring	Proximal radioulnar articulation
	3. Condyloid articulation	Joint cavity. Bony surfaces face shallow convexity, either circular or elliptical	Metacarpophalangeal articulation, wrist joint
	4. Saddle joint	Joint cavity. Bony surfaces reciprocally concavo-convex	First carpometacarpal articulation
	5. Ball and Socket (Enarthrosis)	Joint cavity. Globular head fits into cuplike cavity	Hip joint, shoulder joint
	6. Gliding joint (Arthrodia)	Joint cavity. Opposing bony surfaces planar or nearly so	Carpal or tarsal articulations

EXERCISE 5
COMPARISON OF THE JOINTS OF THE UPPER AND LOWER LIMBS OF MAN

Fill in the following tables from your observations of the upper and lower limbs. Use Table 9-1.

1. **Upper Limbs**

Joint	Class*	Variety	Possible Movements**
Sternoclavicular			
Coracoclavicular			
Acromioclavicular			
Glenohumeral			
Humeroulnar			
Humeroradial			
Proximal radioulnar			
Distal radioulnar			
Intermediate radioulnar			
Radiocarpal			
Intercarpal			
Carpometacarpal (1st)			
Carpometacarpal (2-4)			
Metacarpophalangeal			
Interphalangeal			

*Use the following symbols for joint classes:

F: Freely movable
S: Slightly movable
I: Immovable

**Use the following symbols for joint movements:

Ab: Abduction
C: Circumduction
Ad: Adduction
R: Rotation

2. **Lower Limbs**

Joint	Class*	Variety	Possible Movements**
Sacroiliac			
Symphysis pubis			
Hip joint			
Knee joint			
Superior tibiofibular			
Interosseous tibiofibular			
Inferior tibiofibular			
Ankle joint			
Subtalar			
Talo calcaneonavicular			
Calcaneo cuboid			
Composite joint			
First cuneometatarsal			
Cubo metatarsal			
Metatarso phalangeal			
Interphalangeal			

1. How many freely movable joints are there in the upper limb?

   ____

   How many in the lower limb?

   ____

2. Draw a conclusion regarding the adaptation of each limb for a particular general function.

   Upper limb adapted for: ______________________________

   Lower limb adapted for: ______________________________

LABORATORY 10

Musculature of the Head, Neck, and Trunk

OBJECTIVES

By the end of this laboratory period, you should be able to identify the following.

Muscles

Temporalis
Masseter
Digastric
Mylohyoid
Stylohyoid
Geniohyoid
Sternohyoid
Sternothyroid
Thyrohyoid
Cleidomastoid
Sternomastoid

Nerves

Facial
Massetericus
Deep temporal
Hypoglossal
Spinal accessory

Blood Vessels

Arteries
Common carotid artery
External carotid artery
Internal carotid artery

Veins
Internal jugular vein
External jugular vein
Transverse jugular vein

EXERCISE 1
DISSECTION OF THE HEAD AND NECK

1. Remove the skin and superficial fascia from one side of the face and ventral surface of the neck of the cat.

2. The large **facial veins** will be observed crossing over the fascia covering the **masseter muscle** and converging to form the **external jugular**. The **masseter** is a massive, bulky muscle lying just below and behind the eye. Its three layers of fibers originate from the **malar bone** and insert on the **mandible**. It is an elevator of the **mandible**, and it is innervated by the **massetericus nerve**, one of the seven branches of the **mandibular** division of the **trigeminal nerve**. The **parotid gland** lies in close association with the **masseter** and must be cautiously separated from it before transecting the muscle.

3. The **digastric muscle** is a thick, straplike muscle that lies underneath the **mandible**. It originates on the jugular process of the **occipital bone** and inserts on the ventral border of the **mandible**. When it contracts, it depresses the lower jaw. It is innervated by a branch of the **facial nerve**, which may be traced back to find the **facial nerve** proper. The **mandibular** division of the **trigeminal** also serves this muscle, and a branch, the **inferior alveolar**, can be found at the cranial end.

4. The **temporalis** is another massive muscle that functions in the elevation of the **mandible** in chewing, biting, or carrying objects in the mouth. It is arranged in two layers, the deeper of which originates in the **temporal fossa** and inserts on the internal surface of the **coronoid process** of the **mandible**. The superficial layer originates on the fascia of the **temporal fossa** and inserts on the external surface of the **coronoid process**. This muscle is innervated by the **deep temporal nerve**, a branch of the **mandibular division** of the **trigeminal nerve**.

5. The **mylohyoid** is a thin sheet of muscle lying in the space between the two medial surfaces of the **mandible**. The **digastric** must be transected in order to trace this muscle from its insertion on the median raphe to the origin on the medial surface of the **mandible**. Like the cranial portion of the **digastric**, the **mylohyoid** is innervated by the **inferior alveolar nerve**.

6. The **stylohyoid** is a very slender muscle originating on the **stylohyal bone** and inserting on the **hyoid**. It is an elevator of the **hyoid** and is innervated by the **facial nerve**.

7. Transect the **mylohyoids** by cutting them apart at the median raphe, thus exposing the **geniohyoids**. These narrow muscles originate on the medial surface of the **mandible** near the symphysis and insert on the **hyoid**. They pull the latter bone forward when they contract. Innervation is by way of a branch of the **hypoglossal** nerve.

8. The **external jugular** lies on the surface of the **sternomastoid** and should be carefully dissected away before further muscular dissection.

Figure 10-1. The major head and neck muscles of the cat.

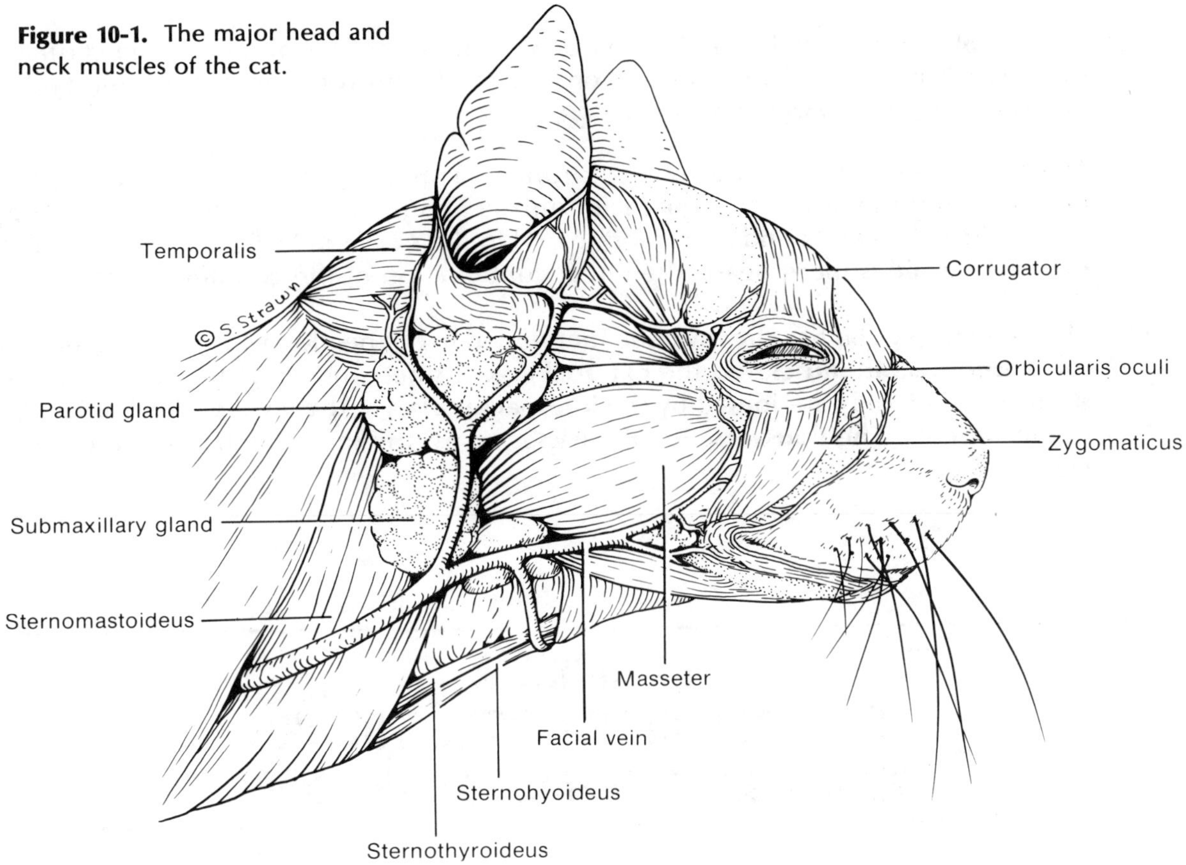

9. The **sternomastoideus** is a wide, flat muscle ascending craniad and laterad as it runs from its origin on the **sternum** to its insertion on the **lambdoidal** ridge and the **mastoid** process of the **temporal** bone. Unilateral contraction of this muscle turns the head to that side and flexes it. Bilateral contraction flexes the head. Innervation is by the **spinal accessory** nerve.

10. Transect the **sternomastoid** and locate the **cleidomastoid** lying beneath it. This rather narrow muscle runs from the **clavicle** to the **mastoid** process of the **temporal** bone. Its action is flexion of the head, if the **clavicle** is fixed, or protraction of the **clavicle** if the head is fixed. Innervation is by the **spinal accessory** nerve and the ventral divisions of C1 and 2.

11. The two narrow bands of muscle on either side of the midline of the throat are the **sternohyoids**. They must be carefully dissected away from the **sternothyroids** which lie underneath. Branches of the **hypoglossal nerve** enter the lateral side of the **sternohyoid.** This muscle originates from the first costal cartilage and inserts on the body of the **hyoid**. The **sternothyroids** share the costal cartilages as a site of origin. In fact, the fibers fuse with those of the **sternohyoid** at that point. However, the deeper muscle inserts on the **caudolateral** surface of the thyroid cartilage of the larynx. It, too, is innervated by the **hypoglossal nerve**.

12. The **thyrohyoid** is a small muscle lying on the lateral border of the **thyroid** cartilage from which it takes origin. It inserts on the caudal horn of the **hyoid** bone and is innervated by the **hypoglossal nerve**.

13. Trace the **common carotid** artery as it runs in a sheath lateral to the **trachea**. The internal carotid splits off from the main trunk at about the level of the medial border of the **digastric muscle** and runs deep to end in the **Circle of Willis** on the brain. The **external carotid** is more nearly the direct continuation of the **common carotid**.

14. The **external jugular** has already been mentioned in connection with the dissection of the **sternomastoid muscle**. As the **external jugulars** descend along the surface of the **sternomastoid**, they are joined by the **internal jugulars** at the level of the cranial end of the **sternum**. The **transverse jugular** passes across the throat high under the lower jaw.

EXERCISE 2
DEEP MUSCLES OF THE TRUNK AND DORSAL SURFACE OF THE NECK

Muscles to be Dissected

Longissimus dorsi
Longissimus capitus
Iliocostalis
Multifidus spinae
Spinalis dorsi
Splenius

Dissection of Trunk Muscles

1. Slit the lumbodorsal fascia from the foot of the tail to a point just caudad of the lowest rib. The **longissimus dorsi** is a long, heavy muscle occupying the gutter formed by the spinous processes and transverse processes of the vertebrae. In the lumbar region, the muscle is separable into a thin medial section and a wide lateral one. In the thoracic region, the two parts fuse together. The dorsal rami of the spinal nerves of the various regions of the cord innervate the **longissimus** as it ascends toward the head. The origin of this muscle is from the **sacrum** and the **sacral** and **lumbar vertebrae**. It inserts along the transverse processes of the lumbar, thoracic, and cervical vertebrae. It is a powerful muscle that acts as an extensor of the spine, a movement of great importance to the cat in running and leaping.

2. Just ventral to the **longissimus dorsi**, several groups of partially separable fibers form the **iliocostalis**. They originate from the **longissimus dorsi** and the angles of the ribs and insert on the ventrolateral surface of the third and fourth ribs. They cause these ribs to pull together when the muscle fibers contract. This muscle or group of muscles is innervated by the dorsal rami of the spinal nerves in the thoracic and lumbar regions.

3. The cranial continuation of the **longissimus dorsi** is a slender, tapering muscle called the **longissimus capitus**. It inserts on the **mastoid process** and is an abductor of the head. It is innervated by the dorsal ramus of C3.

4. Just dorsal and anterior to the **longissimus capitus** is a broad, sheetlike muscle covered by the **clavotrapezius**. This is the **splenius**, originating on the cervical ligament and the fascia of the deeper lying muscles and inserting on the **lambdoidal ridge** of the **occipital** bone. Unilateral contraction of this muscle causes abduction of the head, and bilateral contraction results in neck extension. This muscle is also innervated by the dorsal ramus of C3.

5. The **spinalis dorsi** is a more or less fusiform muscle lying dorsal to the anterior fibers of the **longissimus dorsi**. It originates on the tips of the spinous processes of the ninth or tenth to the thirteenth vertebrae. The fibers run obliquely cranially and laterally to insert on the transverse processes of the first nine thoracic vertebrae and the lower seventh cervical. This muscle is innervated by the dorsal rami of the thoracic nerves.

6. The **multifidus spinae** is made up of bundles of muscle fibers that originate on the transverse processes of the vertebrae and run obliquely, anteriorly, and medially to insert on the spinous processes of vertebrae five or six segments anterior to the origin. In the thoracic region, these fibers are separated from the overlying **longissimus dorsi** with great difficulty. In the cervical region, the fibers form a distinct muscle, the **semispinalis cervicus**, and similarly, in the caudal area they extend posteriorly as the **extensor caudae medialis**. Like the other dorsal muscles, they are innervated by the dorsal rami of the spinal nerves. The **multifidus** and the **spinalis** are extensors of the vertebral column in bilateral contractions and tend to twist or rotate the column with unilateral contractions.

EXERCISE 3
DISCUSSION QUESTIONS

1. Compare the **temporalis** and the **masseter** of the cat in terms of relative size to those of man. How do you account for the difference?

2. What is the difference in function between the deep muscles of the trunk in the cat as compared to man?

LABORATORY 11

Structure of the Circulatory System

A transport system must have some propulsive force that moves it from one locality in the body to another. In addition, it is most efficient if it is enclosed in a set of tubes that carry it to and from the source of its propulsive force. The mechanical requirements, then, of a circulatory system are: first, a pump that provides the propulsive force; second, a system of vessels that carries the transport medium away from the pump to the various parts of the body under sufficient pressure so that it is moved with considerable velocity; third, a system of vessels that permits the exchange of materials between the blood and the surrounding tissues; and finally, a system of tubes that carries the blood back to the pump under much lower pressure.

The mammalian circulatory system beautifully illustrates these requirements. The pump is, obviously, the heart. In fact, the heart is a double pump, one part pumping blood to the body as a whole, the other part to the lungs. The arteries distribute blood to the body. They possess structural features that enable them to withstand great fluctuations in the pressure of the blood as it passes through them. This is especially true of the great elastic arteries, the aorta and the pulmonary artery. Their elasticity, furthermore, is essential to the maintenance of a continuous flow of blood through the system between heartbeats. The arteries branch and rebranch, becoming smaller and smaller in diameter as they probe their way into the tissues of the body. Just before they lose their muscular coat and become capillaries, they are called arterioles.

The vessels whose structure is designed to permit the exchange of materials between the blood and the tissues are the capillaries. They exist in units that permit the shunting of blood from one locality to another, and their thin walls, only one cell in thickness, readily permit respiratory gases, nutrients, waste products, or other materials to pass between the blood and the tissue cells via, of course, the extracellular fluid.

The vessels that return the blood to the heart are the veins. The blood returns under very low pressure because of the high resistance that it encounters in passing through the capillary bed. In addition, it must flow against gravity and against the considerable centrifugal forces that develop as a result of the oscillation of the limbs in locomotion and other movements. The veins of the limbs, therefore, possess valves that prevent back flow of blood and also provide a secondary pumping system in conjunction with the contraction of skeletal muscles surrounding them.

Our purpose in this lab session is to enable you to learn the details of structure of these parts of the circulatory system so that you will have a good foundation for the understanding of their function. The arteries and veins that serve particular portions of the body are studied with that part.

OBJECTIVES

Following this laboratory session, you should be able to identify the following structures on diagrams, on models of the heart, and where clearly visible on specimens of the sheep heart.

1. Aortic arch
2. Innominate artery
3. Left common carotid artery
4. Left subclavian artery
5. Left auricle
6. Left branch of pulmonary artery
7. Apex
8. Superior vena cava
9. Inferior vena cava
10. Right atrium
11. Right atrioventricular valve
12. Right ventricle
13. Myocardium
14. Intraventricular septum
15. Left ventricle
16. Papillary muscles
17. Chordae tendinae
18. Left atrium
19. Left atrioventricular valve
20. Aortic semilunar valve
21. Pulmonary semilunar valve
22. Pulmonary artery
23. Pulmonary vein
24. Innominate artery
25. Left common carotid artery
26. Left subclavian artery

You should also be able to identify the following structures on drawings, photographs, or microscope slides of arteries and veins.

1. Tunica adventitia of artery
2. Tunica adventitia of vein
3. Tunica media of artery
4. Tunica media of vein
5. Tunica intima of artery
6. Tunica intima of vein

MATERIALS

1. Slides and photomicrographs of muscle showing arteries, veins, and capillaries.
2. Sheep heart
3. Model of the circulatory system (demonstration)
4. Model of the heart (demonstration)
5. Corrosion specimens of circulatory systems embedded in plastic

EXERCISE 1
GENERAL SCHEME OF THE CIRCULATORY SYSTEM

Label the following on Figure 11-1.

1. Right atrium
2. Left atrium
3. Right ventricle
4. Left ventricle
5. Aorta
6. Pulmonary artery
7. Pulmonary veins
8. Inferior vena cava
9. Superior vena cava
10. Circulation to head and arms
11. Circulation to viscera
12. Circulation to legs
13. Pulmonary circulation

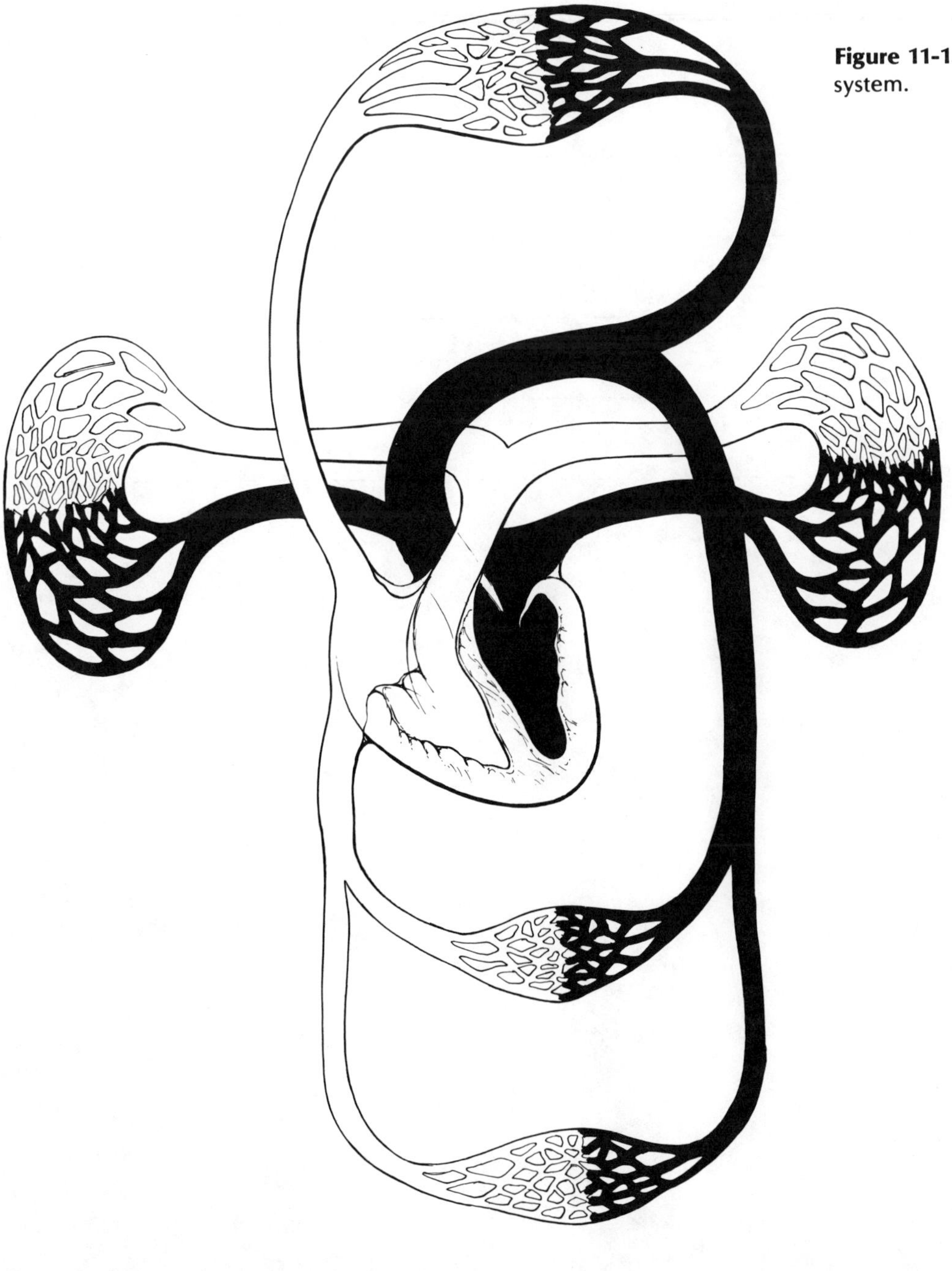

Figure 11-1. Scheme of the circulatory system.

EXERCISE 2
STRUCTURE OF THE HEART

Examine the heart model on the demonstration table, and dissect the sheep heart to locate as many as possible of the structures listed on Table 11-1. Label the structures on Figures 11-3 and 11-4, and list the function of each.

Table 11-1

Structure	**Function**
1. Aortic arch	__________
2. Innominate artery	__________
3. Left common carotid artery	__________
4. Left subclavian artery	__________
5. Left auricle	__________
6. Left branch of pulmonary artery	__________
7. Apex	landmark__________
8. Superior vena cava	__________
9. Inferior vena cava	__________

THE HUMAN SYSTEMS

INDEX

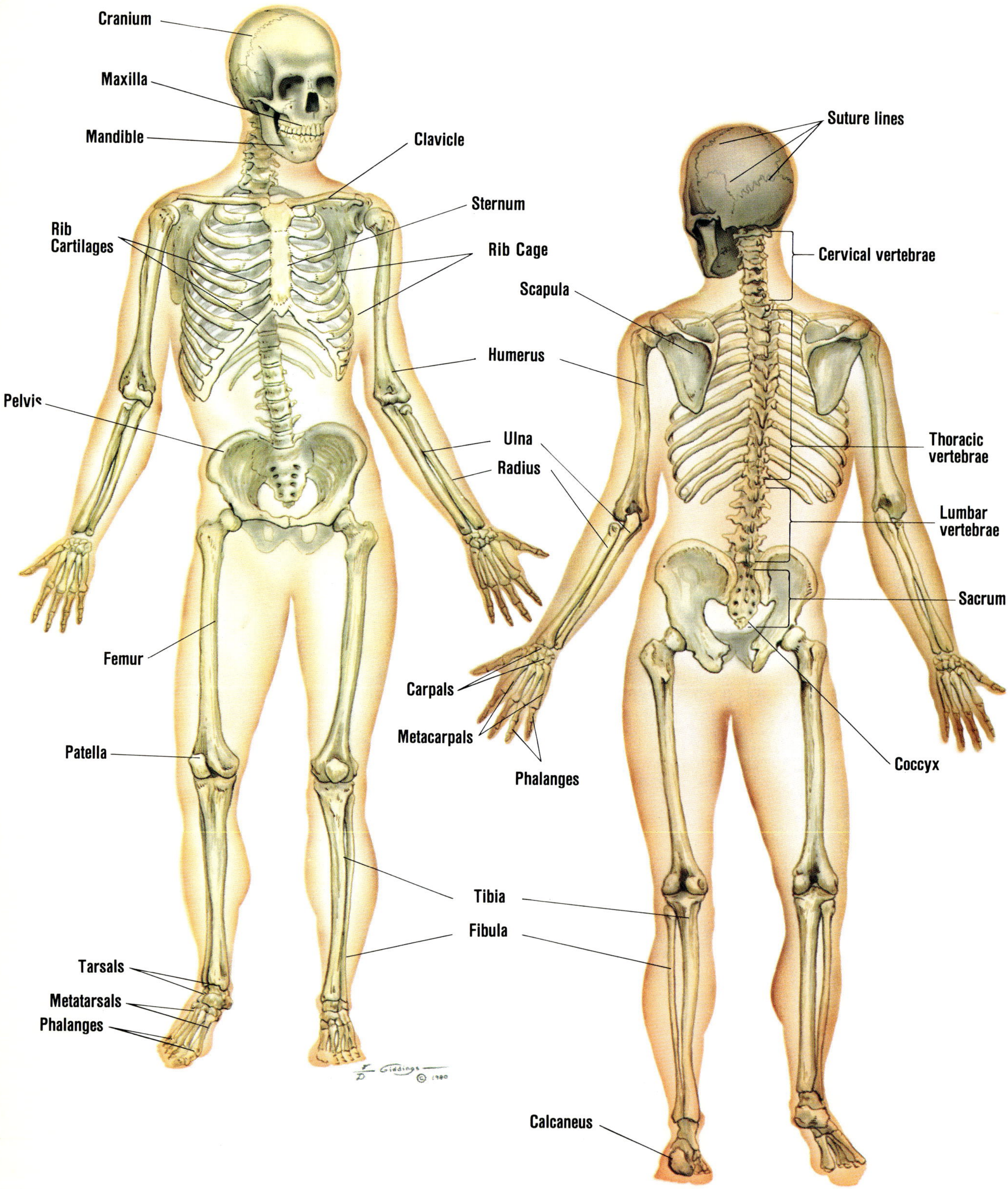

THE SKELETAL SYSTEM

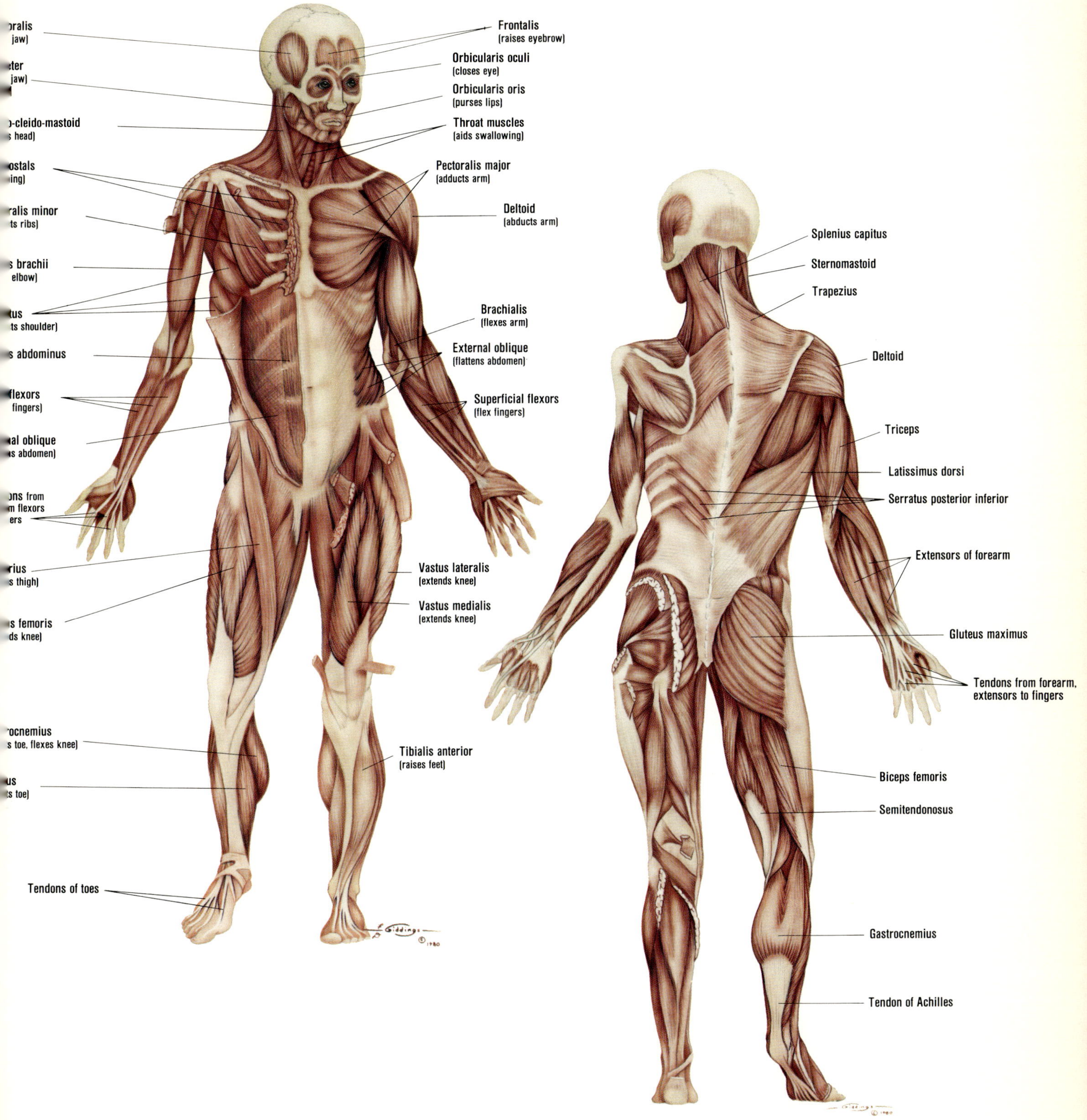

THE MUSCULAR SYSTEM

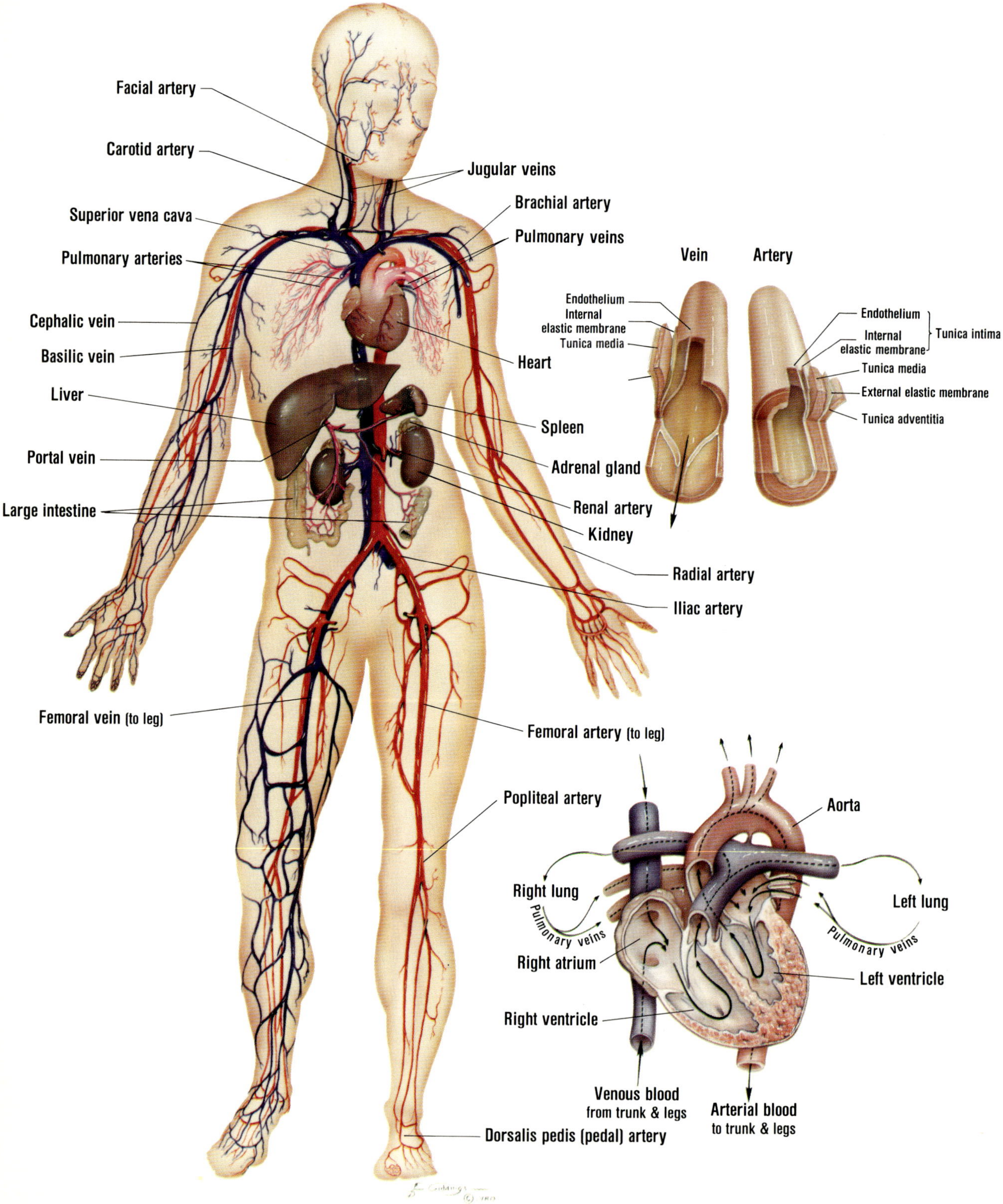

THE CIRCULATORY SYSTEM

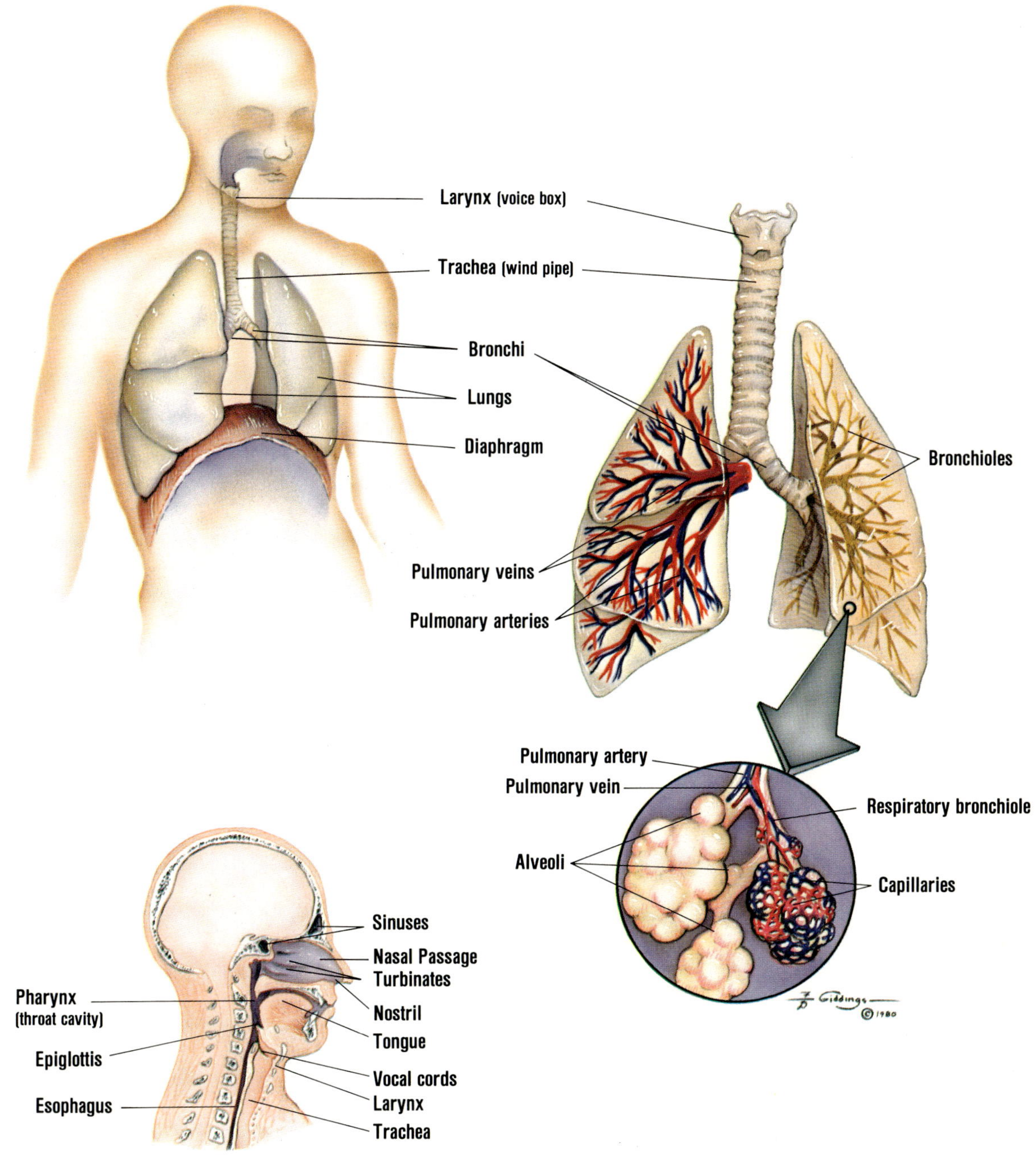

THE RESPIRATORY SYSTEM

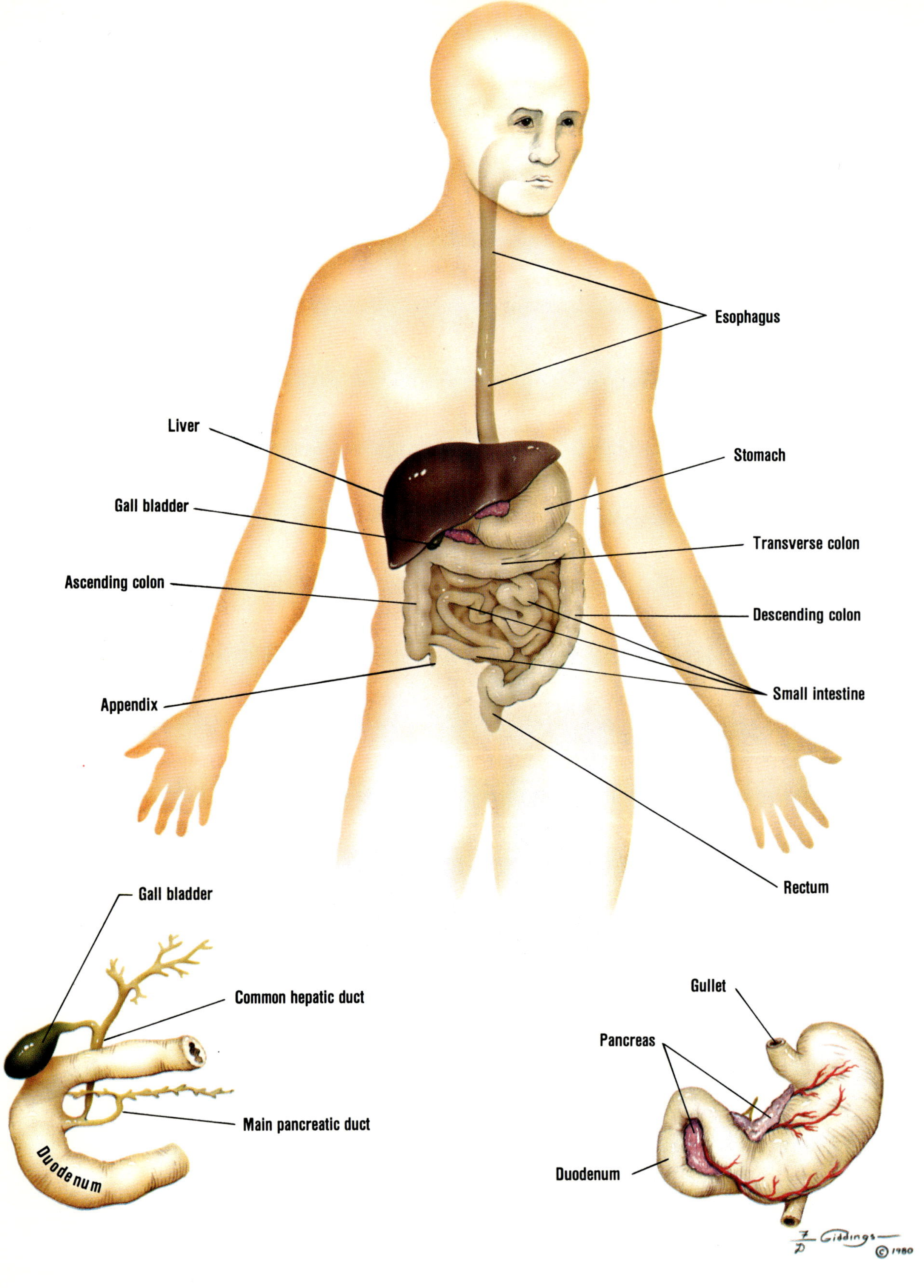

THE DIGESTIVE SYSTEM

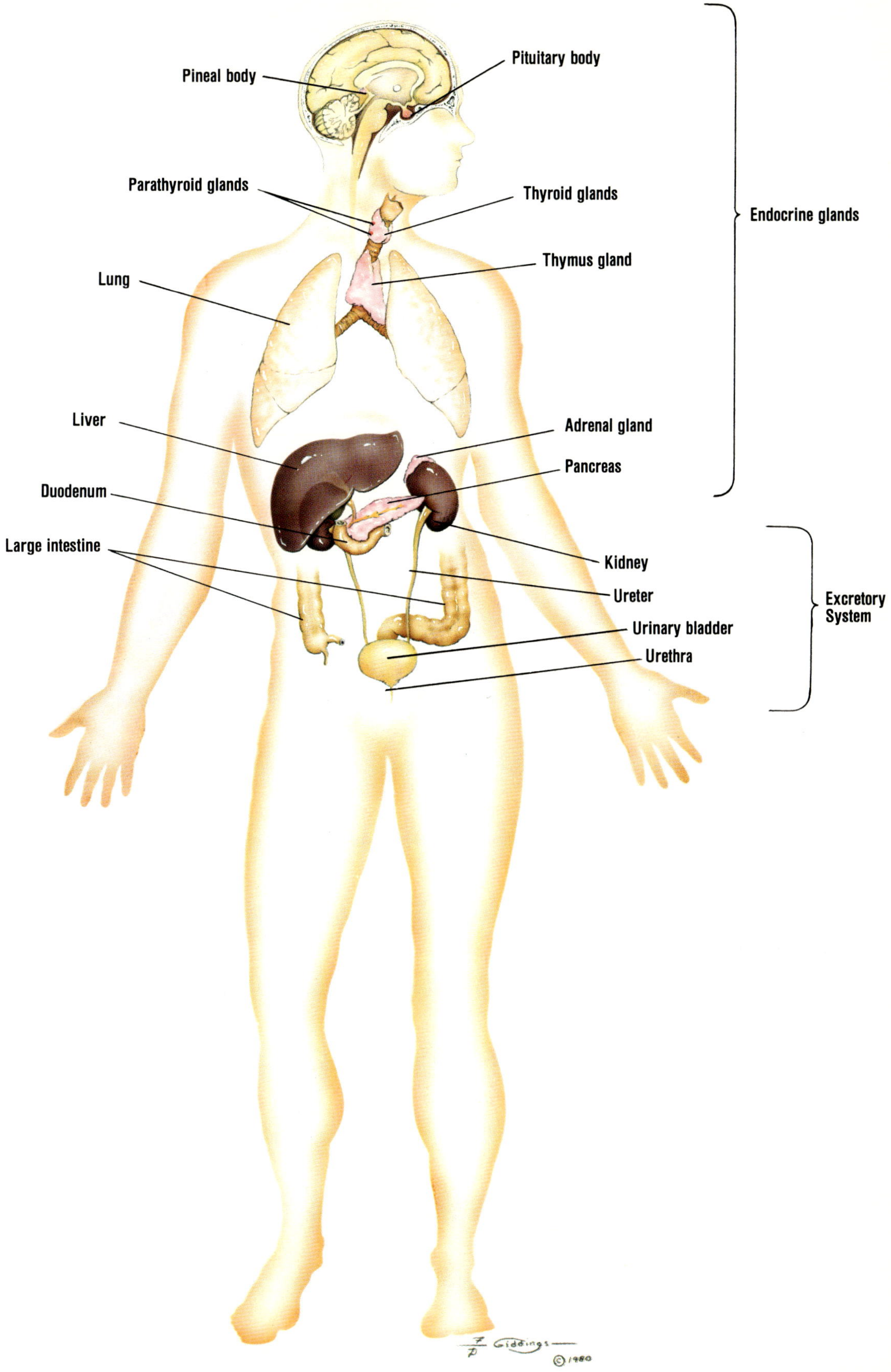

THE URINARY AND ENDOCRINE SYSTEM

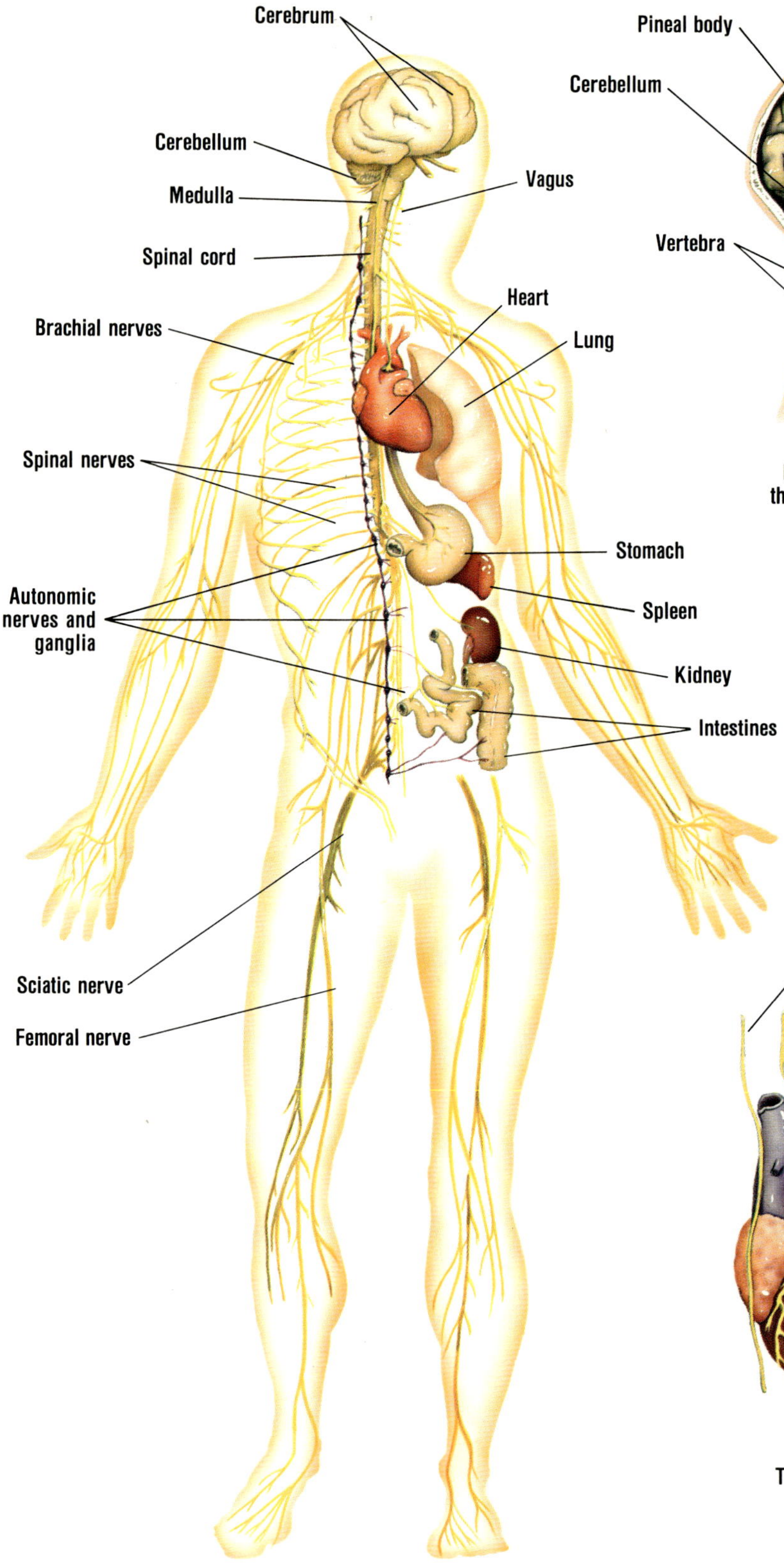

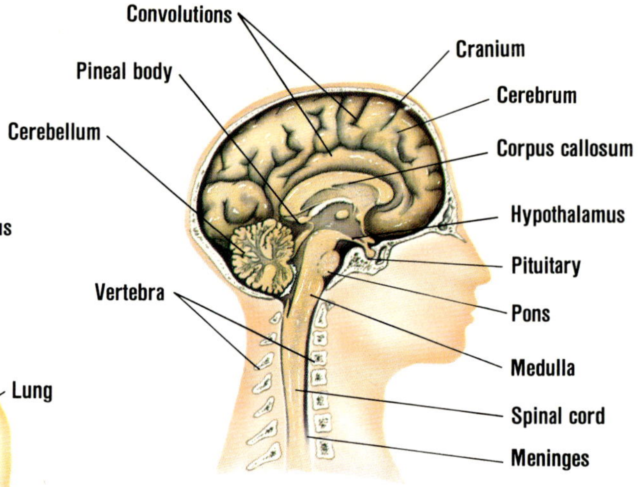

Midsaggital Section of
the brain and spinal cord

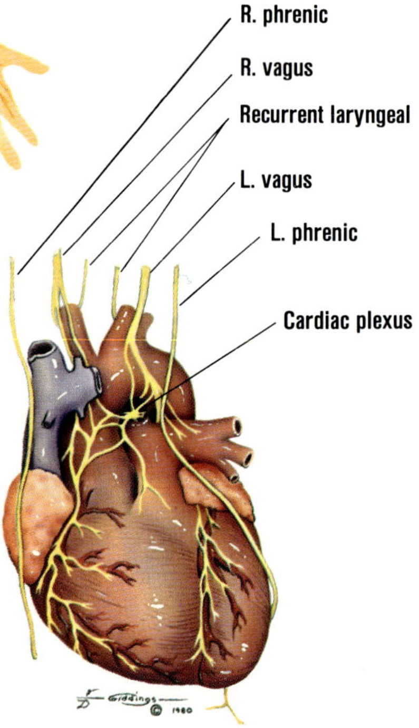

The Nerves of the Heart

THE NERVOUS SYSTEM

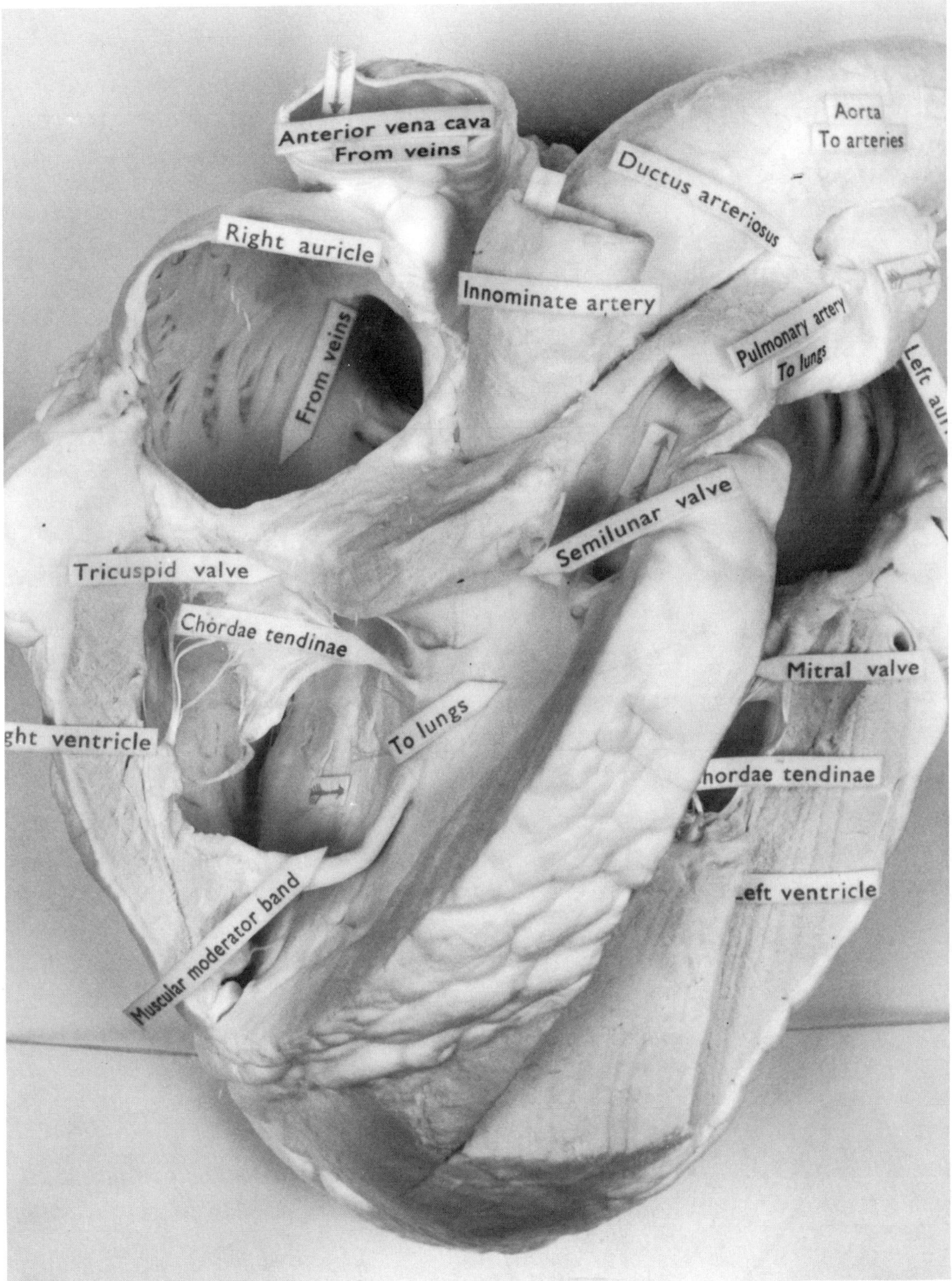

Figure 11-2. Sheep heart. Alternative terminology includes right and left atrium instead of right and left auricle, right atrio-ventricular valve instead of tricuspid valve, and left atrio-ventricular instead of mitral valve.

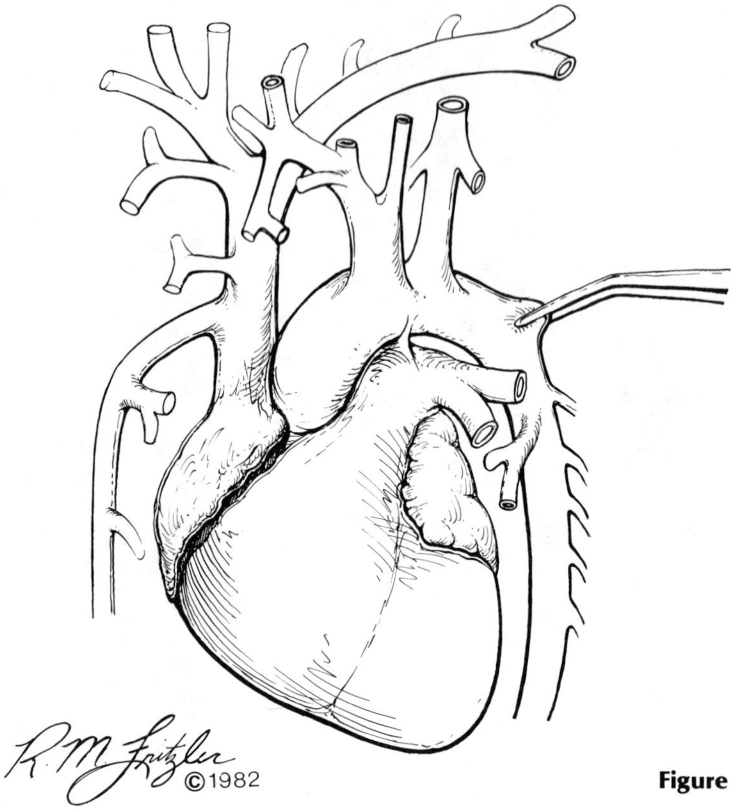

Figure 11-3. The cat heart and great vessels.

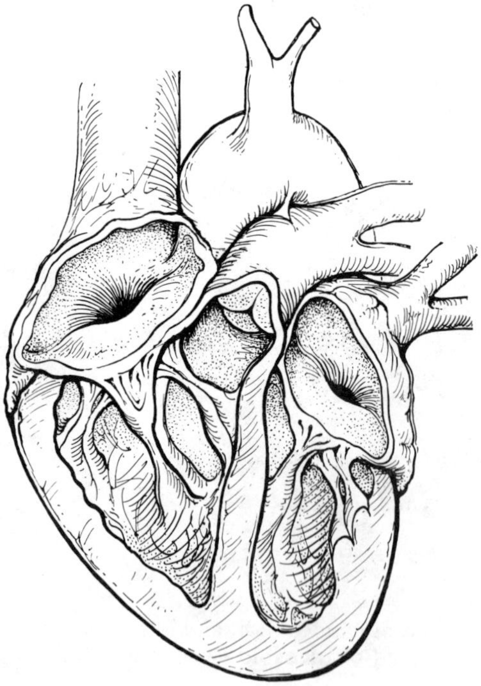

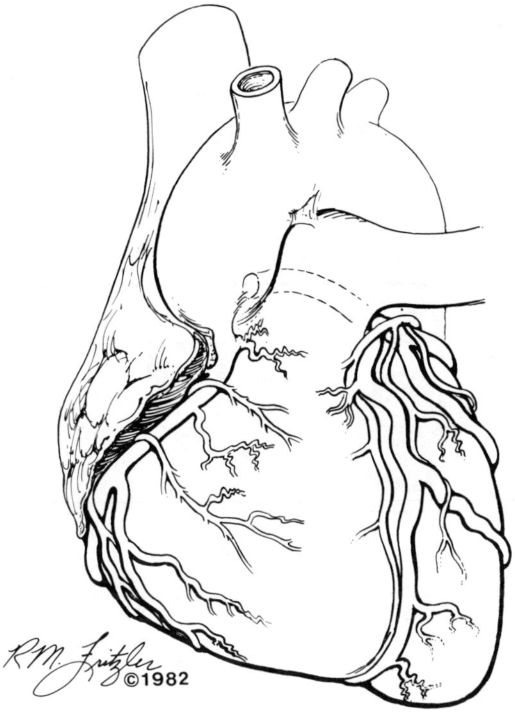

Figure 11-4. Gross anatomy of the human heart, dorsal and ventral views.

Ventral

Dorsal

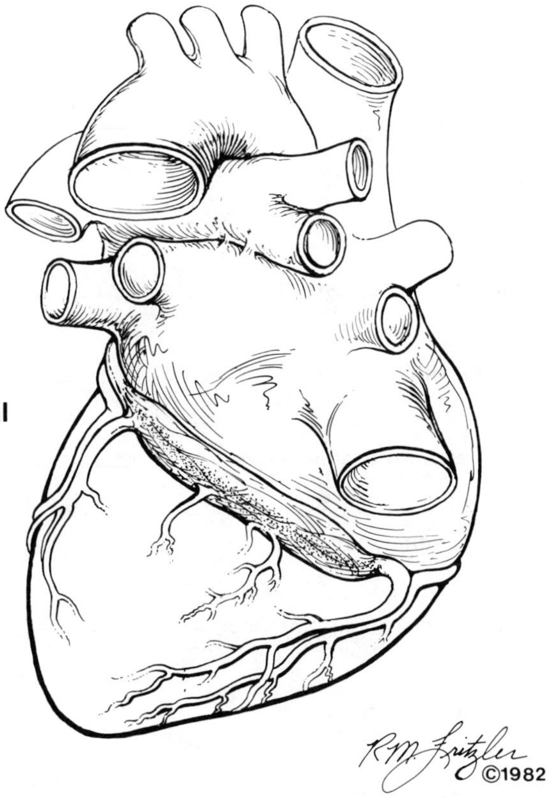

List the function of each of the following structures.

Table 11-2

Structure	Function
1. Superior vena cava	______
2. Inferior vena cava	______
3. Right atrium	______
4. Right atrioventricular valve	______
5. Right ventricle	______
6. Myocardium	______
7. Intraventricular septum	______
8. Left ventricle	______
9. Papillary muscles	______
10. Chordae tendineae	______
11. Left atrium	______
12. Left atrioventricular valve	______
13. Aortic semilunar valve	______
14. Pulmonary semilunar valve	______
15. Pulmonary artery	______
16. Pulmonary vein	______

17. Innominate artery	____________
18. Left common carotid artery	____________
19. Left subclavian artery	____________

EXERCISE 3
THE ARTERIES AND VEINS

On Figure 11-5, label the following structures, and in Table 11-3, indicate the function.

Table 11-3

Structure	Function
1. Tunica adventitia	____________
2. Tunica media	____________
3. Tunica intima	____________

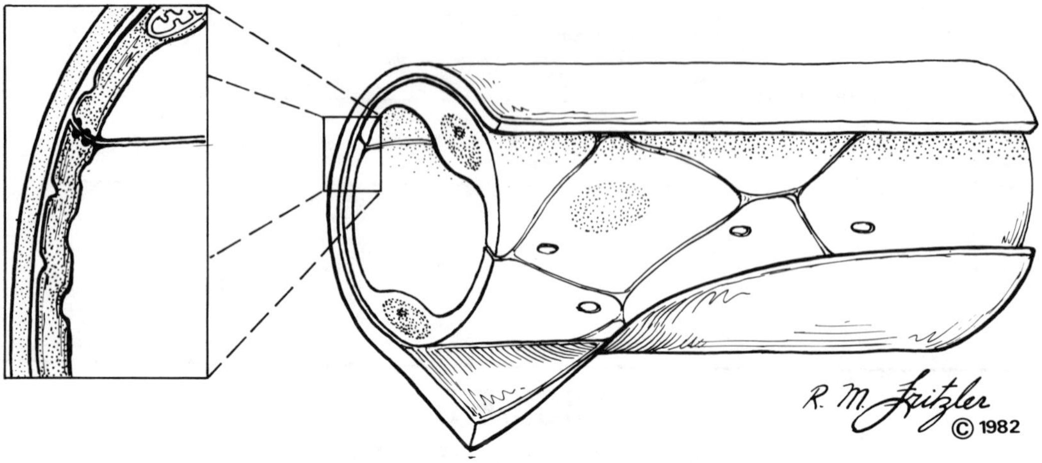

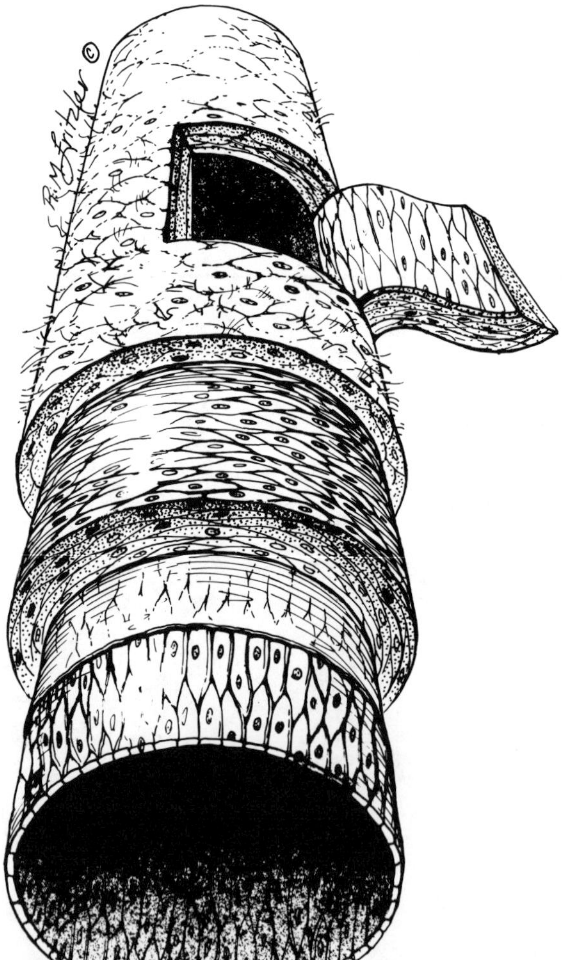

Figure 11-5. Three-dimensional drawing of the coats of blood vessels. Top view shows a capillary; bottom view shows a medium-sized artery. With the aid of your instructor, label the following: 1—adventitia; 2—media; 3—intima.

Figure 11-6. Arterial system of the cat. Locate and label the following arteries: 1. Aortic arch; 2. Thoracic aorta; 3. Abdominal aorta; 4. Left subclavian a.; 5. Right subclavian a.; 6. Left common carotid a.; 7. Right common carotid a.; 8. Left and right brachial a.; 9. Left and right renal a.; 10. Left and right common. iliac a.

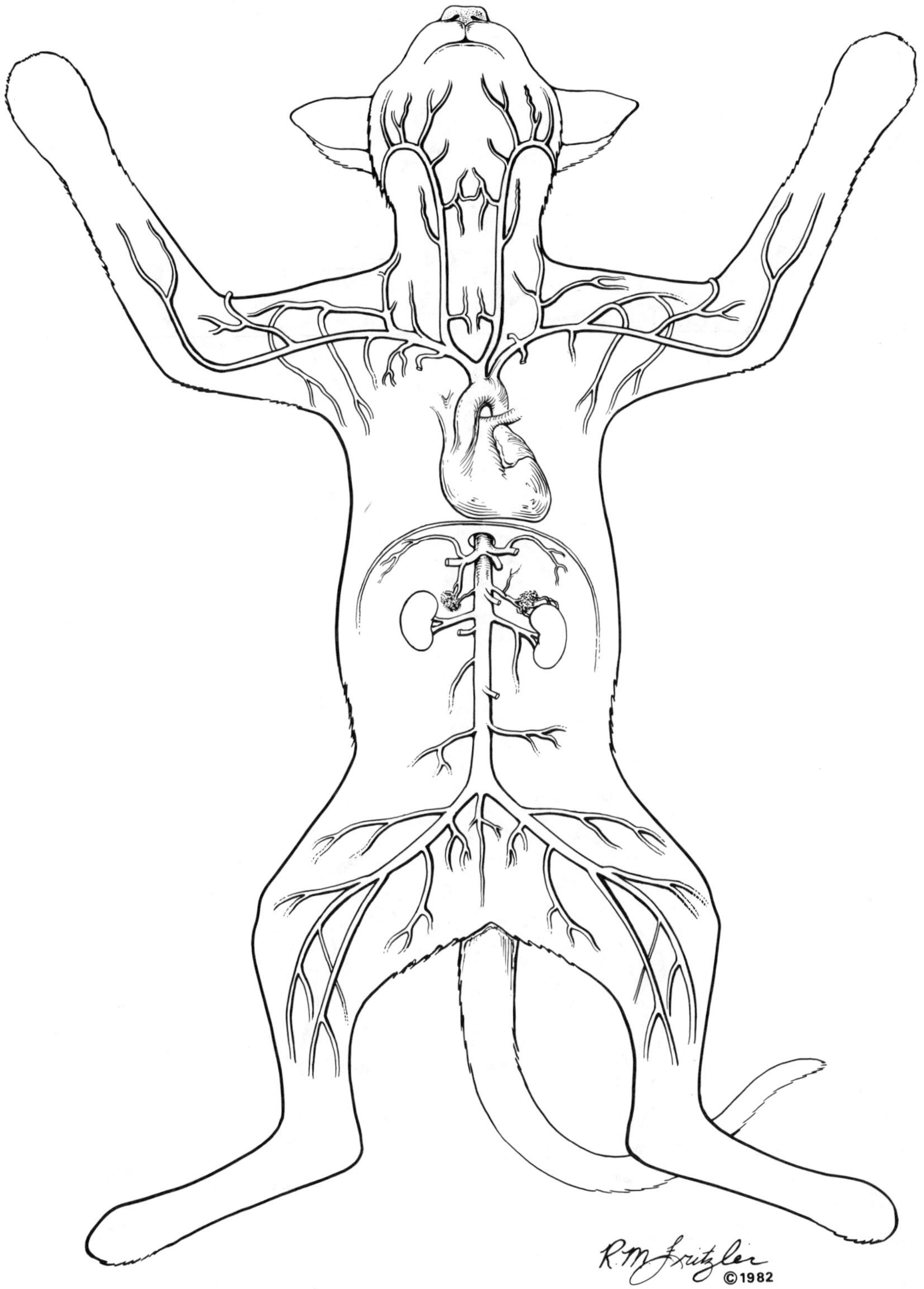

Figure 11-7. Venous system of the cat. Locate and label the following veins: 1. Superior vena cava; 2. Inferior vena cava; 3. Left and right jugular v.; 4. Right and left subclavian v.; 5. Right and left brachial v.; 6. Right and left renal v.; 7. Right and left common iliac.

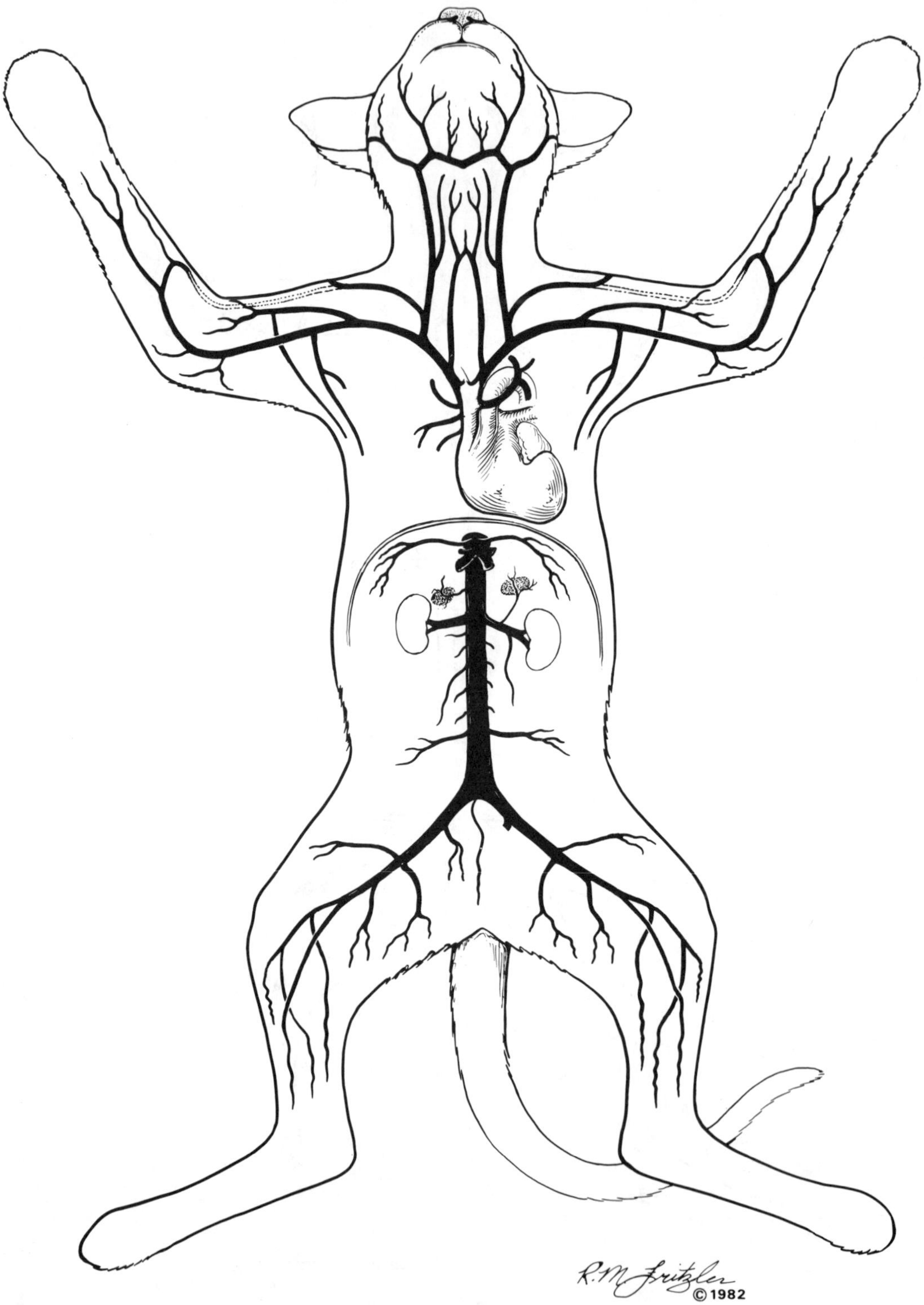

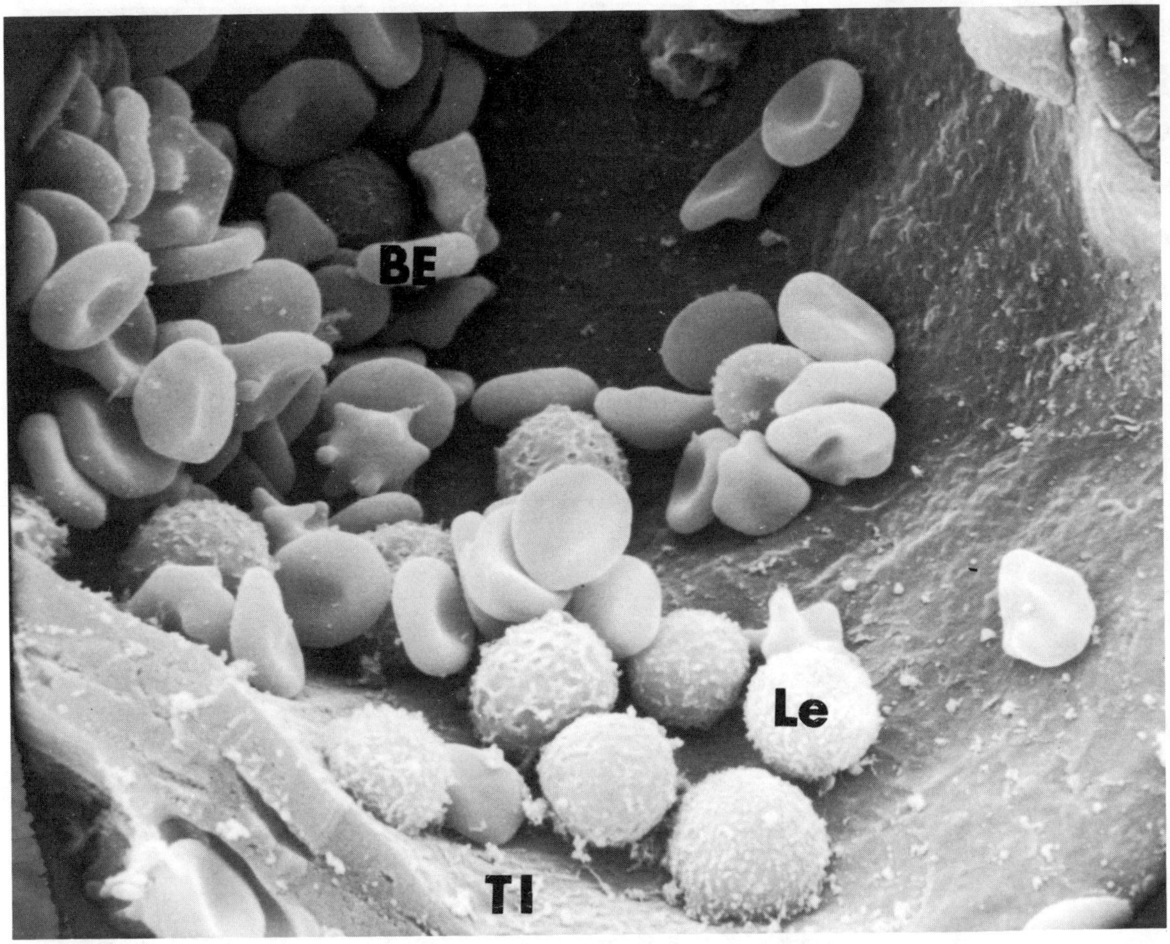

Figure 11-8. Scanning electron micrograph of blood cells, 613X. BE—erythrocytes; Le—leukocytes; Ti–tunica intima.

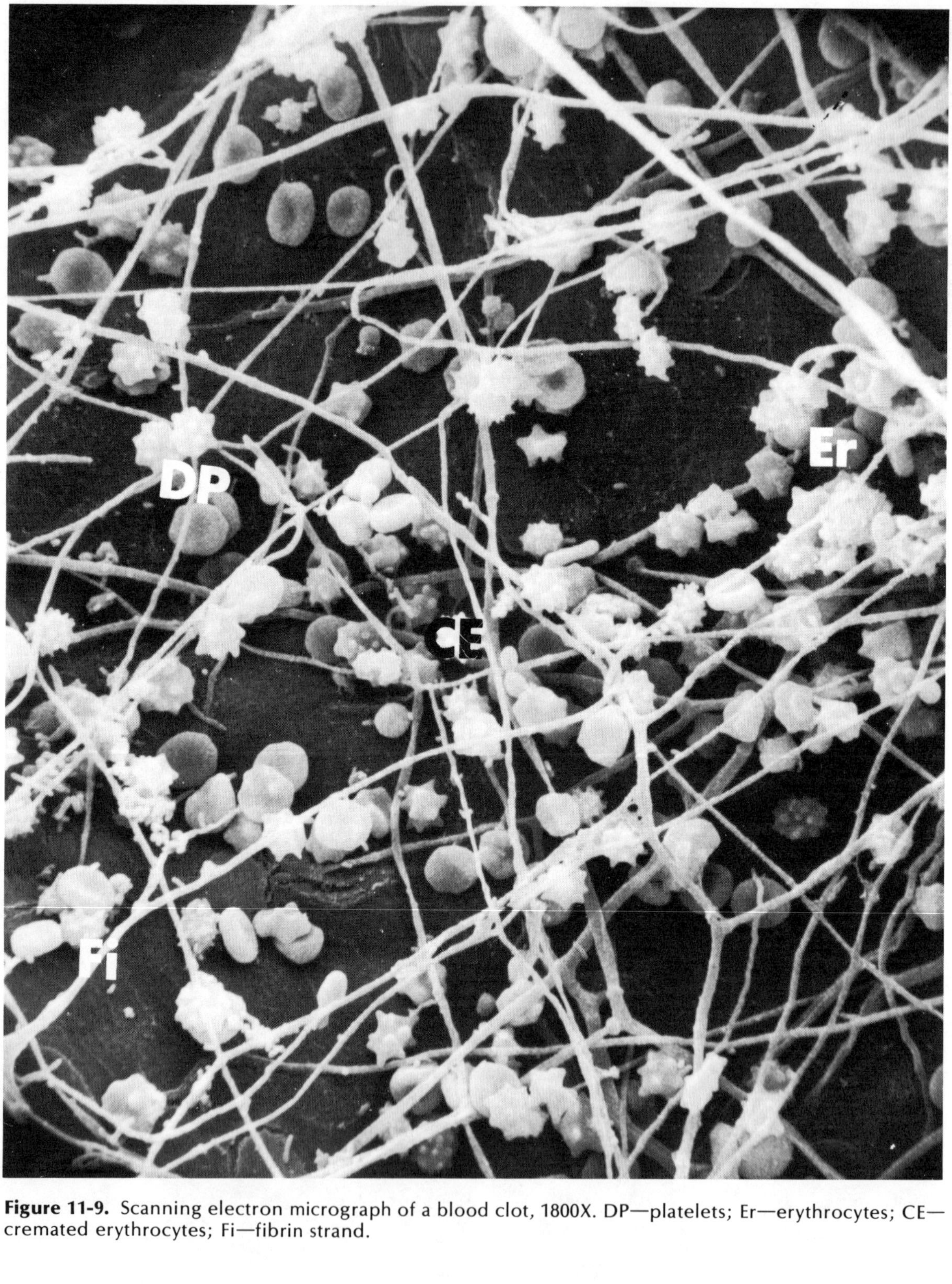

Figure 11-9. Scanning electron micrograph of a blood clot, 1800X. DP—platelets; Er—erythrocytes; CE—cremated erythrocytes; Fi—fibrin strand.

LABORATORY 12

Structure of the Respiratory System

OBJECTIVES

This laboratory session is to enable you to locate and identify the following structures of the cat respiratory system.

Upper Respiratory Tract

Nasal cavity
Nasopharynx
Larynx, including the thyroid and cricoid cartilages, epiglottis, glottis, and vocal chords
Trachea, including cartilage rings
Primary bronchi

Lower Respiratory Tract

Lung lobes
Secondary bronchi

Thorax

Ribs
Sternum
Internal and external intercostal muscles
Diaphragm
Parietal pleura
Visceral pleura
Vagus nerve
Phrenic nerve
Intercostal nerves, arteries, and veins

Also, you should be able to identify the following structures on microscope slides, photomicrographs, drawings, or diagrams and be able to state their functional significance.

Bronchioles
Alveolar sacs
Arteries, veins, and capillaries
Alveoli
Alveolar sacs
Alveolar ducts

EXERCISE 1
DISSECTION OF THE UPPER RESPIRATORY TRACT OF THE CAT

1. Slit the musculature and skin of the cheek on both sides so that the lower jaw can be reflected downward to expose the entire **pharynx**. If necessary, cut the **mandible** with a bone saw.

2. The nasal portion of the **pharynx** in the cat lies above the rather extensive membranous fold of the **soft palate**. Observe the boundaries and communication of the **nasal pharynx**.

3. Carefully transect the musculature of the floor of the mouth, and locate the **hyoid bone**. Separate the **larynx** from the **esophagus**, and cut through the trachea about 1 cm below the posterior section of the **larynx**. Remove the **larynx** and hyoid bone, and carefully clear the connective tissue so that the muscles and cartilages can be observed. Locate the large, shield-shaped **thyroid** cartilage, the **cricoid**, and the two **arytenoids**. Also find the **epiglottis**, a large, leaf-shaped cartilage covering the opening at the anterior aspect of the **larynx**, the **glottis**.

4. Slit the **larynx** by making an incision through the posterior wall. Observe the false vocal cords, which are two folds of mucous membrane running ventrodorsally from the **thyroid** to the **arytenoid** cartilages. These structures are not involved in phonation but serve to shut the inspired air in the lungs in straining. The true vocal cords lie posterior to the false cords and extend from the **thyroid** cartilage to the **arytenoids**.

5. Trace the nerves of the laryngeal muscles back to the vagus.

The Thoracic Wall

1. **Scalenes**

   The middle **scalene** is the longest and most easily identified of this muscle group. It lies

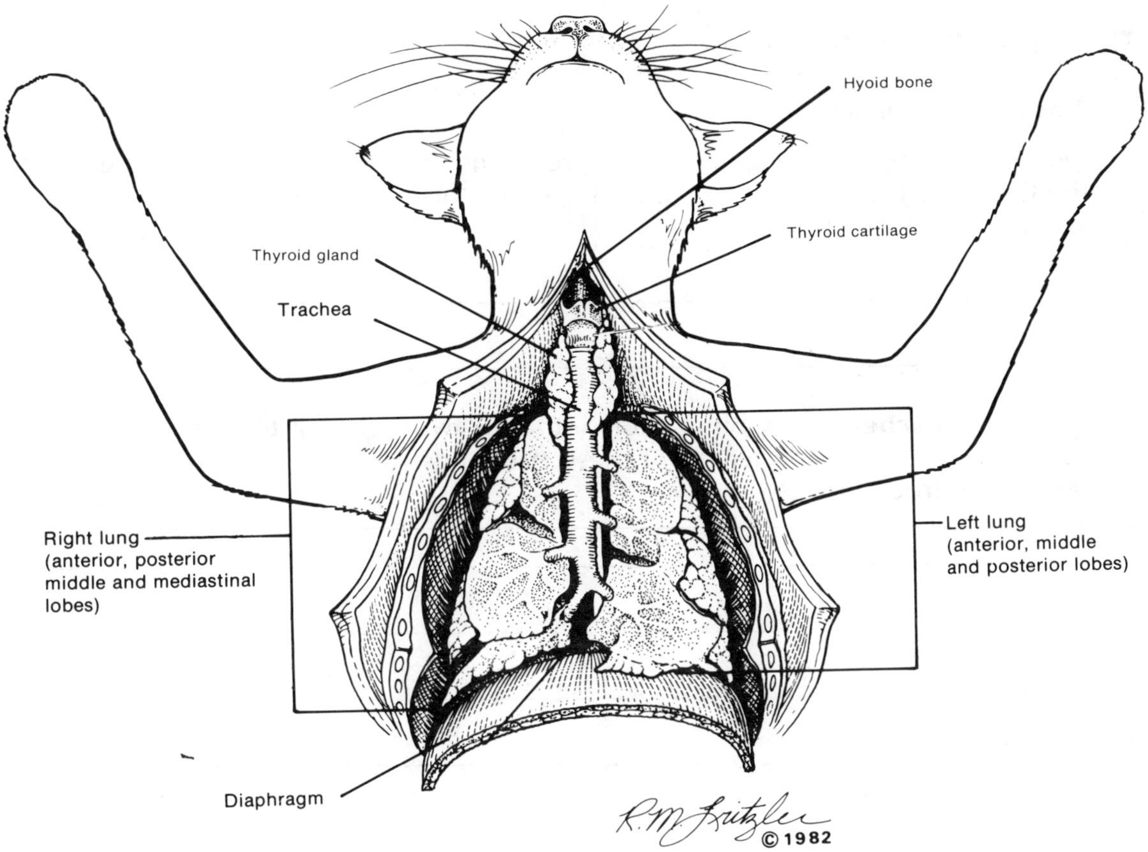

Figure 12-1. Respiratory system of the cat.

over the **thoracic** wall an inch to the side of the midline. The origin of the **serratus ventralis** lies just lateral and underneath the middle **scalene**. The **dorsal** and **ventral scalenes** are located at the cranial end of the middle **scalene** and respectively lateral and medial to it. These muscles flex the neck or pull the ribs cranially. They are innervated by the ventral rami of the first to sixth cervical nerves.

2. Make a longitudinal incision through the thoracic wall, starting above the first rib about three-fourths inch from the midline and continuing down to the insertion of the **diaphragm**. Extend the incision laterally to both sides. Cut the ribs at the ventral border of the **iliocostalis** muscle, and turn the thoracic wall laterally. Observe the following structures.

   a. **Intercostals**

   Peel off the connective tissue covering one of the **external intercostal** muscles and note fiber direction, origin, and insertion. Dissect away the parietal pleura from the **internal intercostals** and locate an **intercostal nerve, artery**, and **vein**.

   b. **Diaphragm**

   Trace the attachments of the **diaphragm**, and observe the structures that penetrate it. Find the **phrenic** nerve at its point of entrance into the **diaphragm**, and follow it **cranially** up the side of the **posterior vena cava** into the **mediastinum**.

The Lower Respiratory Tract

1. **Trachea and Primary Bronchi**

   Observe the location, length, diameter, relations, and structure of the **trachea**. Notice the C-shaped incomplete rings of cartilage that form the framework of this structure. What is their function?

   ____

   ____

2. Trace the **trachea** to its bifurcation into the two **primary bronchi**. Observe the diameter, direction, and length of these structures. How do they compare with the trachea in structure?

   ____

   ____

   ____

   ____

3. **Lungs**

   Examine the shape, pleural investment, lobes, and fissures of the lungs. Insert a straw into the **trachea**, and fold the membranous wall tightly around the straw. Blow forcibly into the straw, and inflate the lungs of the cat. Observe the collapse of the lungs when pressure is removed from the straw.

   a. What can you infer regarding the elasticity of the lungs from this experiment?

      ____

   b. Dissect the root of the lung. Identify the **pulmonary arteries** and **veins**, and the **secondary bronchi**. How many secondary bronchi are there?

      ____

      ____

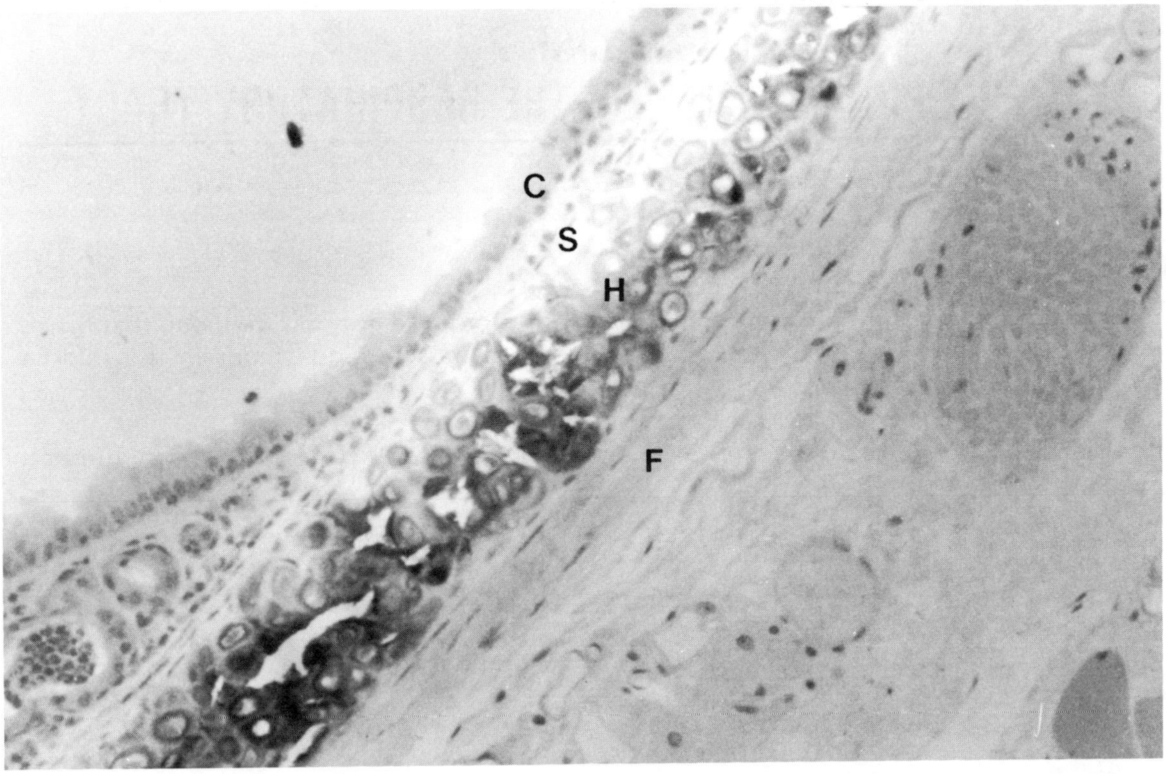

Figure 12-2. Low-power photomicrograph of mouse trachea. C—ciliated columnar epithelium; S—submucosa; H—hyalin cartilage; F—fibrous connective tissue.

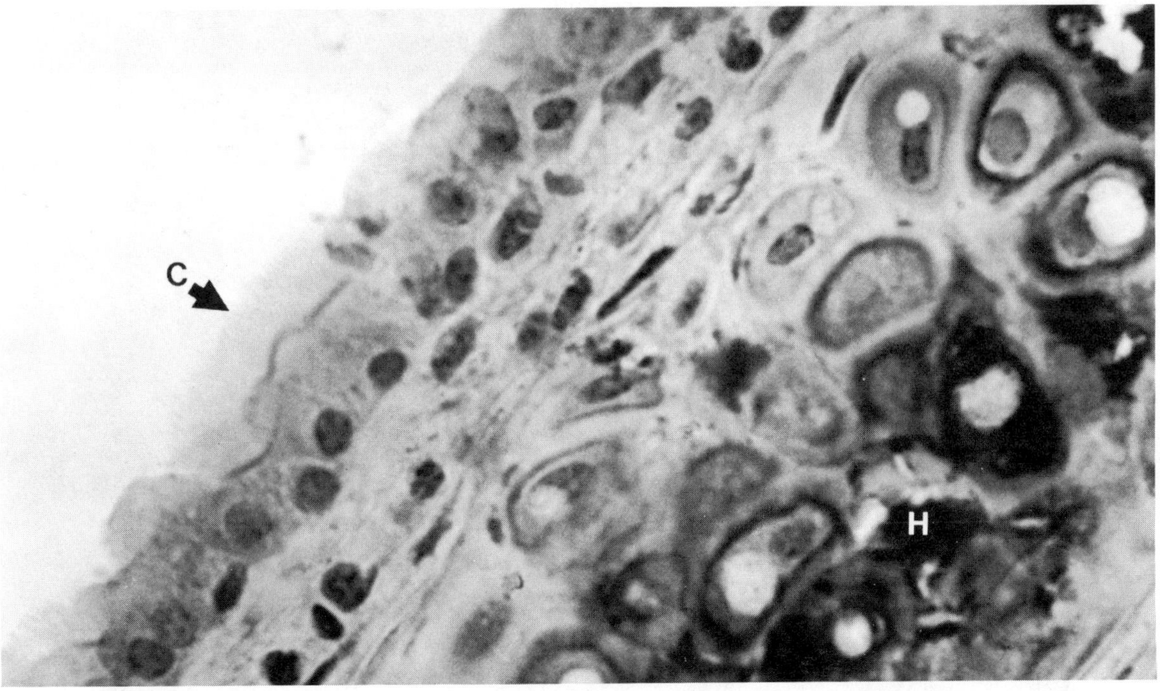

Figure 12-3. High-power photomicrograph of mouse trachea showing: C—cilia on columnar epithelial cells; H—hyalin cartilage.

EXERCISE 2
MICROSCOPIC ANATOMY OF THE RESPIRATORY TRACT

Trachea

1. Examine prepared slides of the trachea. Locate the dense connective tissue that forms the outer layer; the cartilagenous rings, smooth muscle, and glands in the middle layer; and the ciliated columnar epithelium lining the trachea.

2. Draw a diagram of the cross section of the trachea, including the structures mentioned above.

3. Draw four or five ciliated columnar epithelial cells and goblet cells showing a typical arrangement.

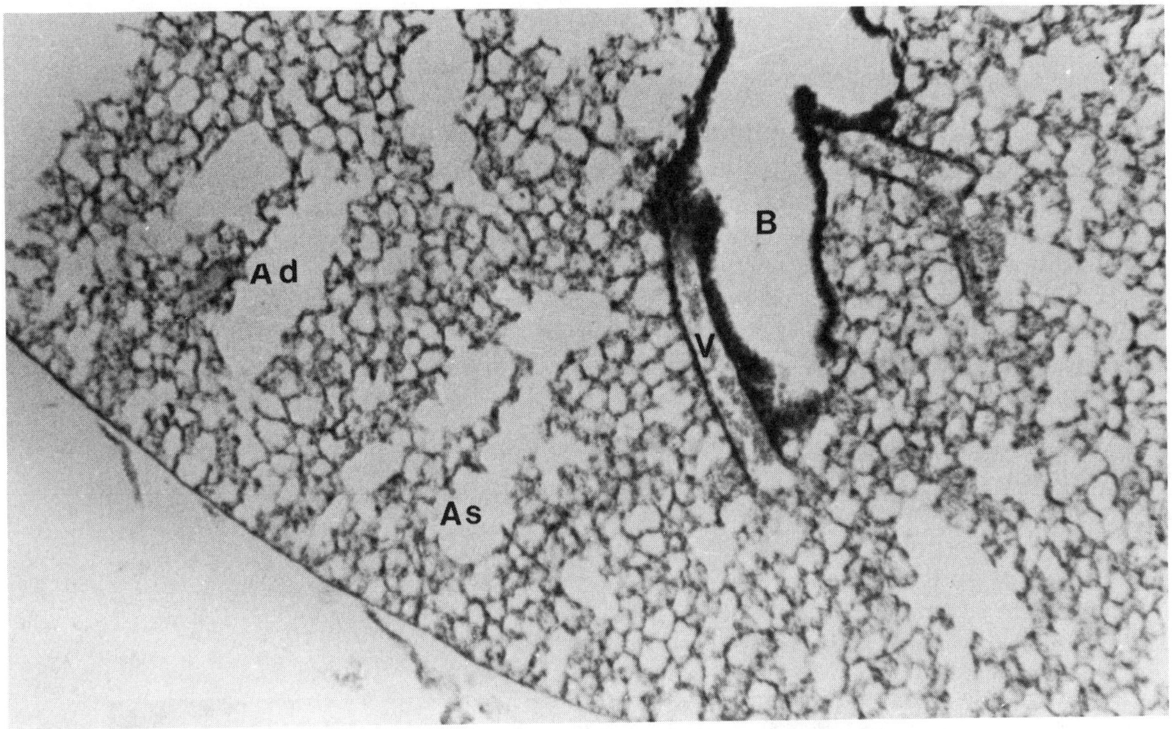

Figure 12-4. Lung 50X.

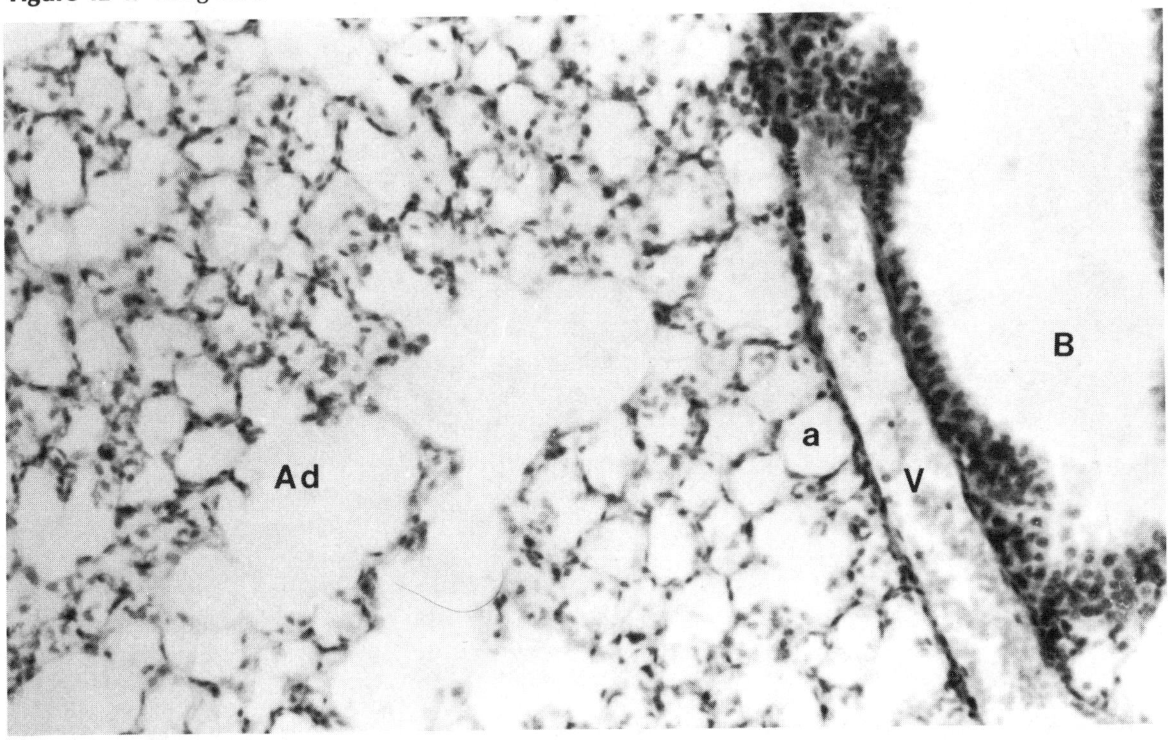

Figure 12-5. Lung 200X. B—bronchiole; As—alveolar sac; Ad—alveolar duct; V—vein (note red blood corpuscles); A—alveolus.

Lung

Study prepared slides of the lung, and locate the following structures:

1. Bronchiole: Observe the epithelial lining, the smooth muscle, the elastic connective tissue, and any plaques of hyalin cartilage that may be present.

2. Alveolar sacs and alveoli: Identify the central alveolar duct and the thin-walled squamous epithelial cells forming the walls of the alveolar. Locate a capillary by searching for a clump of red blood cells.

3. On the accompanying drawing of the ultrastructure of the alveolar capillary complex, label:

   Nucleus of endothelial cells
   Cytoplasm of endothelial cell
   Endothelial basement membrane
   Red blood corpuscles
   Nucleus of epithelial cell
   Cytoplasm of epithelial cell
   Epithelial basement membrane

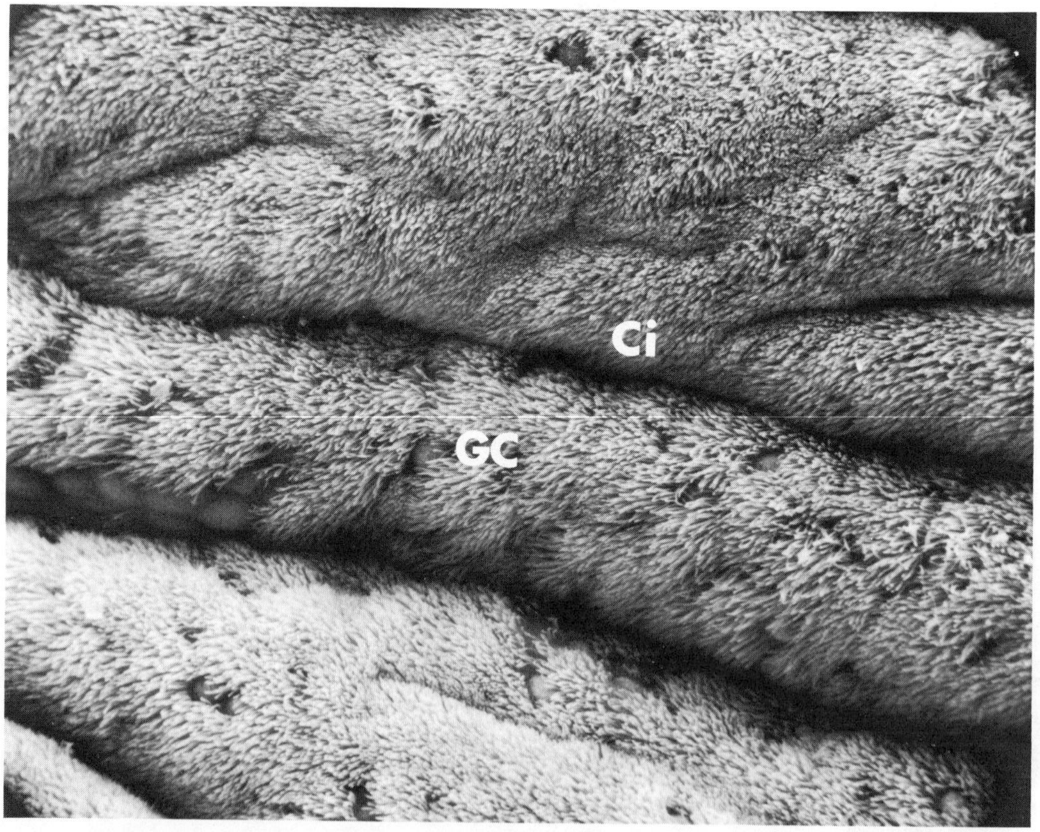

Figure 12-6. Scanning electron micrograph of the lining of the trachea, 1310X. CI—cilia; GC—goblet cells.

Figure 12-7. Ultrastructure of the alveolar capillary complex.

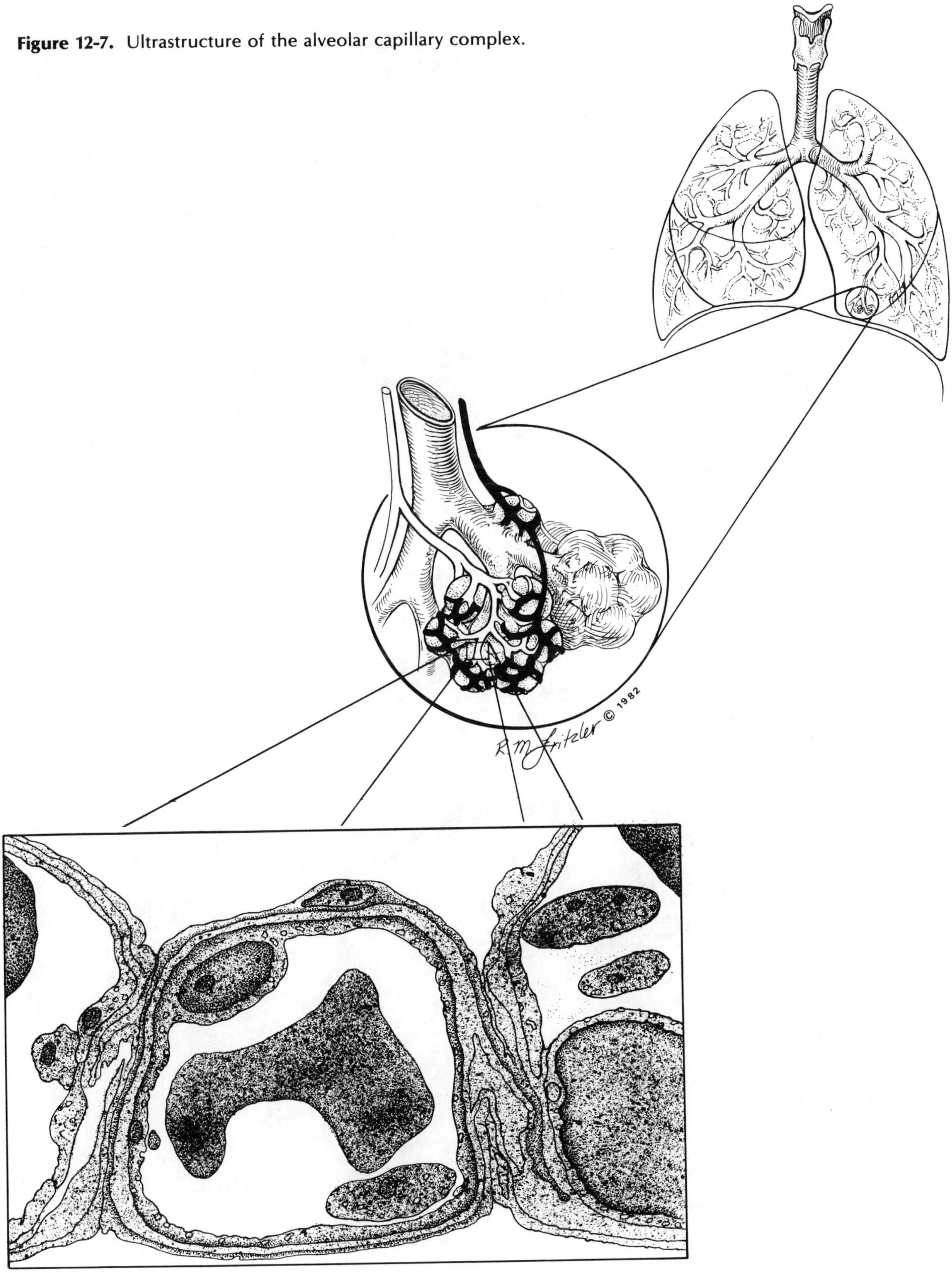

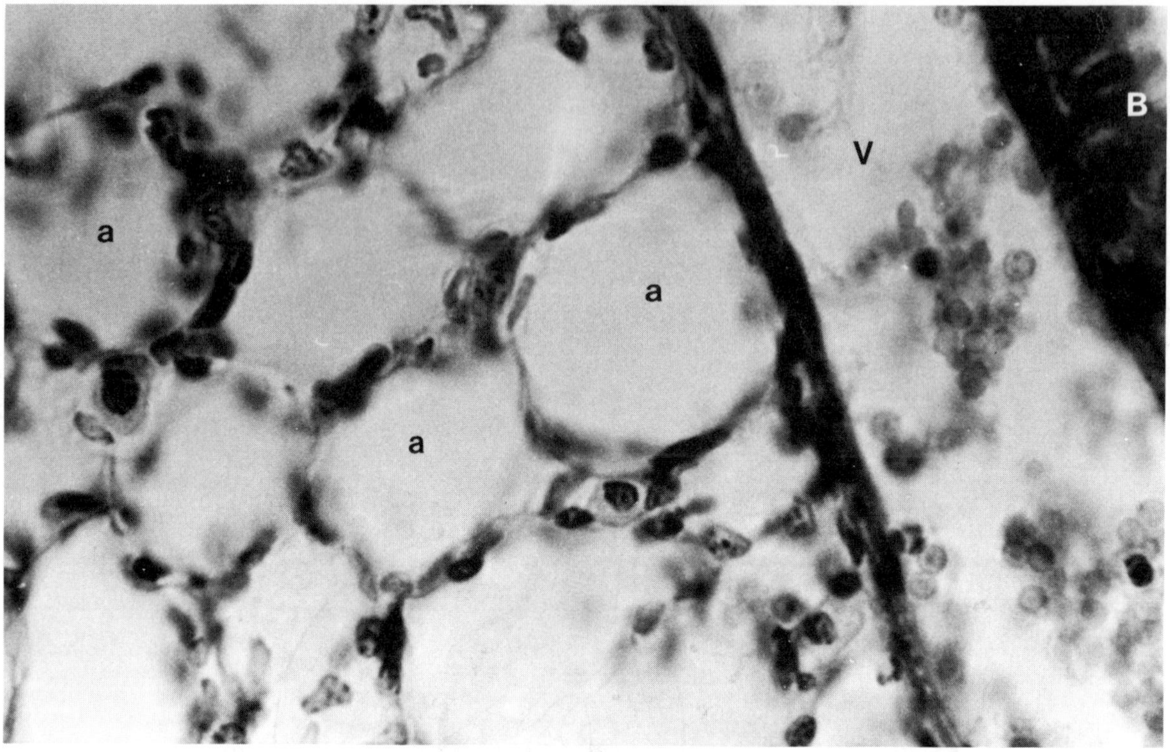

Figure 12-8. Lung 500X. B—bronchiole (note columnar epithelial lining); V—vein (note red blood corpuscles); a—alveolus (note variety of cell types in walls).

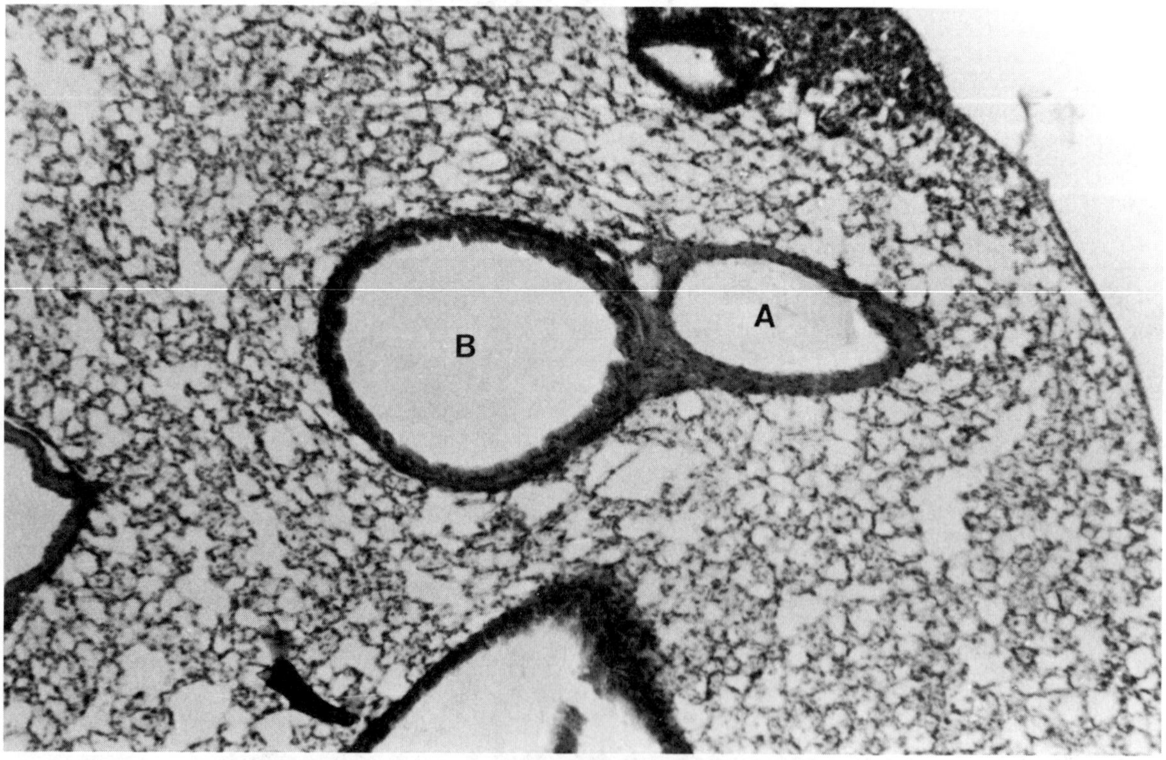

Figure 12-9. Lung 50X. A—artery; B—bronchiole.

LABORATORY 13

Structure of the Digestive Tract

OBJECTIVES

After finishing this laboratory session, you should be able to identify the following parts of the digestive system.

The Coats

Identify on diagrams and photographs or in a cat dissection the following coats of the digestive tract, and describe or recognize a description of their differences in different parts of the tract.

1. Serosa
2. Muscularis
3. Mucosa

Subdivisions of the Digestive Tract

Identify on diagrams, models, or a cat dissection the following subdivisions of the digestive tract and the structures listed. Be able to describe or recognize a description of their function.

1. **Oral cavity**—tongue, teeth, oral pharynx
2. **Esophagus**
3. **Stomach**—cardiac sphincter, pyloric sphincter, rugae, fundus, body, pylorus
4. **Small Intestine**—duodenum, villi, jejunum, plica circulares, ileum, cecum, duodenal papilla
5. **Large Intestine**—ascending colon, hepatic flexure, transverse colon, splenic flexure, descending colon, sigmoid flexure

Accessory Glands

Identify on diagrams or a cat dissection, and describe or recognize a correct description of the function of the following organs, including the substance secreted and the control of the secretory process when appropriate.

1. **Parotid gland** and **Stenson's duct**
2. **Liver**—cystic duct, hepatic duct, common bile duct
3. **Gall bladder**
4. **Pancreas**—pancreatic duct

EXERCISE 1
DISSECTION OF UPPER DIGESTIVE TRACT OF THE CAT

The Mouth

The mouth is an ovoid cavity subdivided into the **vestibule,** between the lips and cheeks, and the oral cavity, which is bounded ventrally by the tongue, laterally by the teeth, and dorsally by the palate.

1. Observe the hard and soft palates. The former is formed by the palatine processes of the **maxillae** and **pre-maxillae** and the palatine bones. The oral surface is covered with a thick mucose containing many corrugations called the transverse palatine ridges. The soft palate is a long, musculo-membranous sheet that separates the caudal portion of the **oral cavity** from the **nasal cavity**. It runs from the caudal edge of the **hard palate** to the level of the **epiglottis.**

2. Differentiate between the **incisors, canines, premolars,** and **molars.** Fill in Table 13-1, showing the number of teeth in the upper and lower half jaws. Refer to Figures 4-3 and 4-5 of the cat skull.

Table 13-1

	Incisors	Canines	Premolars	Molars
Upper Right				
Lower Right				

How many teeth does the cat have in its entire jaw, and how many in its lower?

How do the teeth of the cat differ from those of man?

____

3. Locate the following structures of the **tongue:**

   a. **Filiform papillae**

      Spiny projections on the dorsal surface toward the tip.

   b. **Fungiform papillae**

      Broad and rounding knobs that taper to a narrow stalk, where they are attached to the surface of the tongue. These papillae have taste buds in their epithelium.

   c. **Circumvallate papillae**

      Flat, round structures with deep trenches surrounding them. They form two or three rows in a V-shaped pattern pointing toward the base of the tongue. They also possess taste buds in their epithelium.

      1. What are the functions of the tongue in the cat?

         ____

         ____

      2. How are the various types of papillae related to these functions?

         ____

         ____

4. **Salivary glands**

   In addition to the three salivary glands found in man, the **parotid**, **sublingual**, and **submandibular**, the cat possesses two others, a **molar** and an **infraorbital**.

   a. The parotid is the largest of the salivary glands and lies just ventral to the **pinna** of the external ear, with its cranial border lying over the caudal border of the **masseter** muscle. Its duct runs across the surface of the **masseter** to its cranial border, where it turns in, penetrating the **buccinator**, and opens the largest cusp of the third premolar. In some specimens, small masses of glandular tissue lie along the duct. These are called **accessory parotids**.

   b. The **submandibular** gland lies ventral to the **parotid** and under the angle of the **mandible**. It is usually less lobulated than the **parotid** and is surrounded by two large lymph glands. The duct emerges from the internal surface and runs across the lateral border of the **digastric** muscle. It enters the oral cavity from a large papilla, which lies lateral to the cranial end of the **frenulum** (a fold of mucous membrane that connects the cranial end of the tongue to the floor of the mouth).

c. The **sublingual** gland is a small, tooth-shaped structure that lies along the duct of the **submandibular**. Its duct follows the duct of the **submandibular** and opens into the same papillae. However, whereas the duct of the **submandibular** gland opens on the lateral surface of the papilla, that of the **sublingual** opens on the medial surface.

5. **Pharynx** and **Esophagus**

   a. Observe the boundaries and communications of the nasopharynx and oropharynx.

   b. Study the **esophagus**, noting its length, course, and penetration of the **diaphragm**. Slit this tube open and examine the mucosa for glands.

EXERCISE 2
DISSECTION OF THE ABDOMEN

The Abdominal Cavity

1. Make a midline incision through the skin of the abdomen from the **sternum** to the **pubic symphysis**.

2. Identify the **rectus abdominis**—two thin, flat strips of muscle about one centimeter wide lying on either side of the midline raphe. Locate the **external oblique**, and make a longitudinal incision through it parallel to the lateral margin of the **rectus abdominis**. Peel out the **internal oblique** and the **transversus** from the inner surface of the abdominal wall. Note the parietal peritoneum.

3. Reflect the muscular walls laterally, and observe the **greater omentum** covering the abdominal contents, like an apron. Notice the extensive fat deposits in this saclike structure. Carefully trace the origin.

4. The lesser omentum is a double fold of peritoneum extending caudally from the liver to the lesser curvature of the **stomach.** Identify this structure and find the foramen of Winslow, or the **epiploic foramen**, which lies dorsal to the free border of the **lesser sac** and the **greater sac**, which line the two subdivisions of the peritoneal cavity.

The Abdominal Viscera

The Liver

Note the location, size, color, and texture of the liver. Are you able to deduce its function from its structure? Find the **falciform** (sickle-shaped) ligament, a double fold of peritoneum that suspends the liver from the diaphragm. Locate the **gall bladder** on the inferior

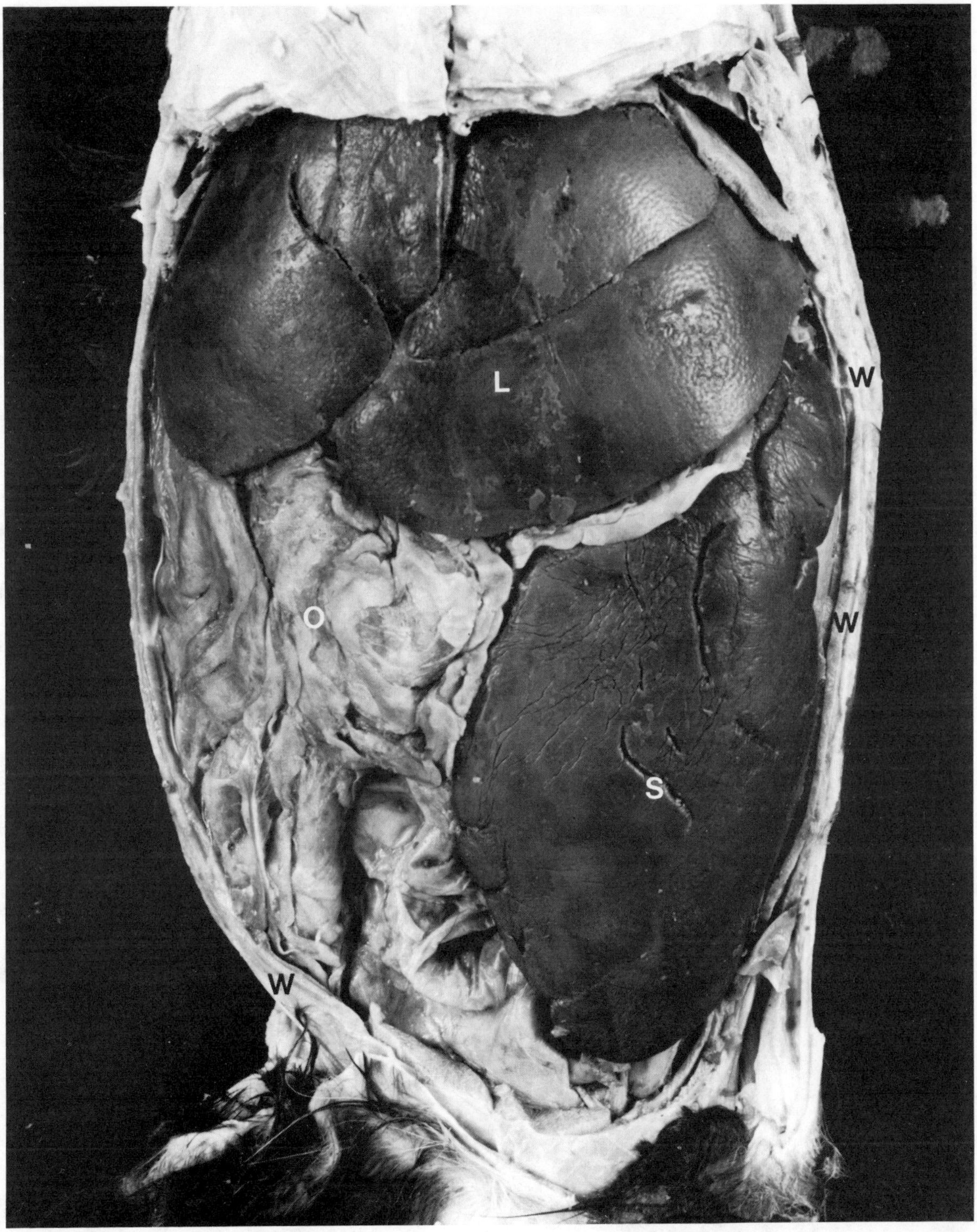

Figure 13-1. Abdominal cavity. L—liver; W—abdominal wall; S—spleen (greatly enlarged); O—greater omentum.

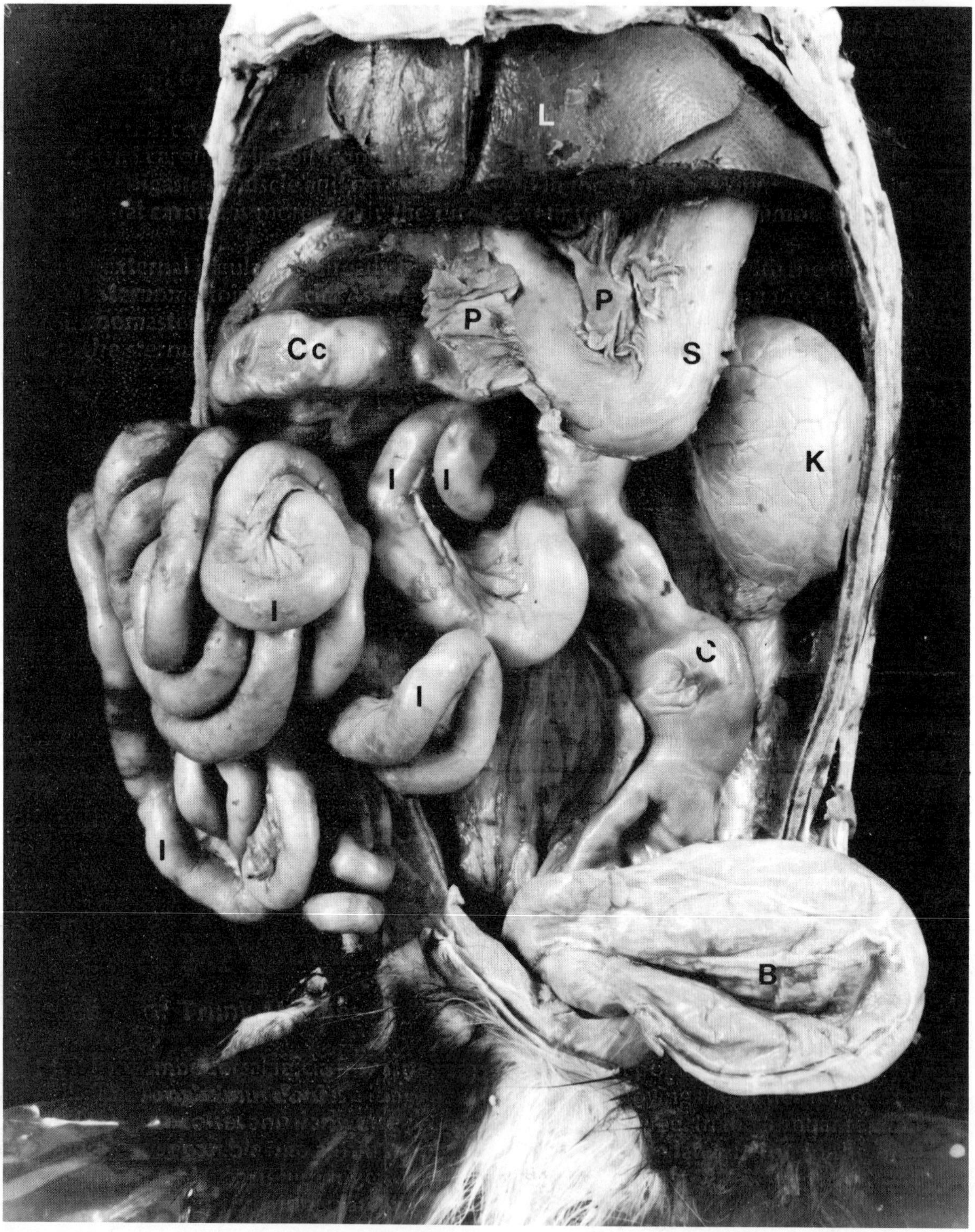

Figure 13-2. Abdominal cavity, greater omentum, spleen, and part of liver removed. L—liver; K—kidney; S—stomach; I—small intestine; C—colon; Cc—cecum; B—urinary bladder; P—pancreas (part removed with greater omentum).

Figure 13-3. Abdominal viscera of the cat.

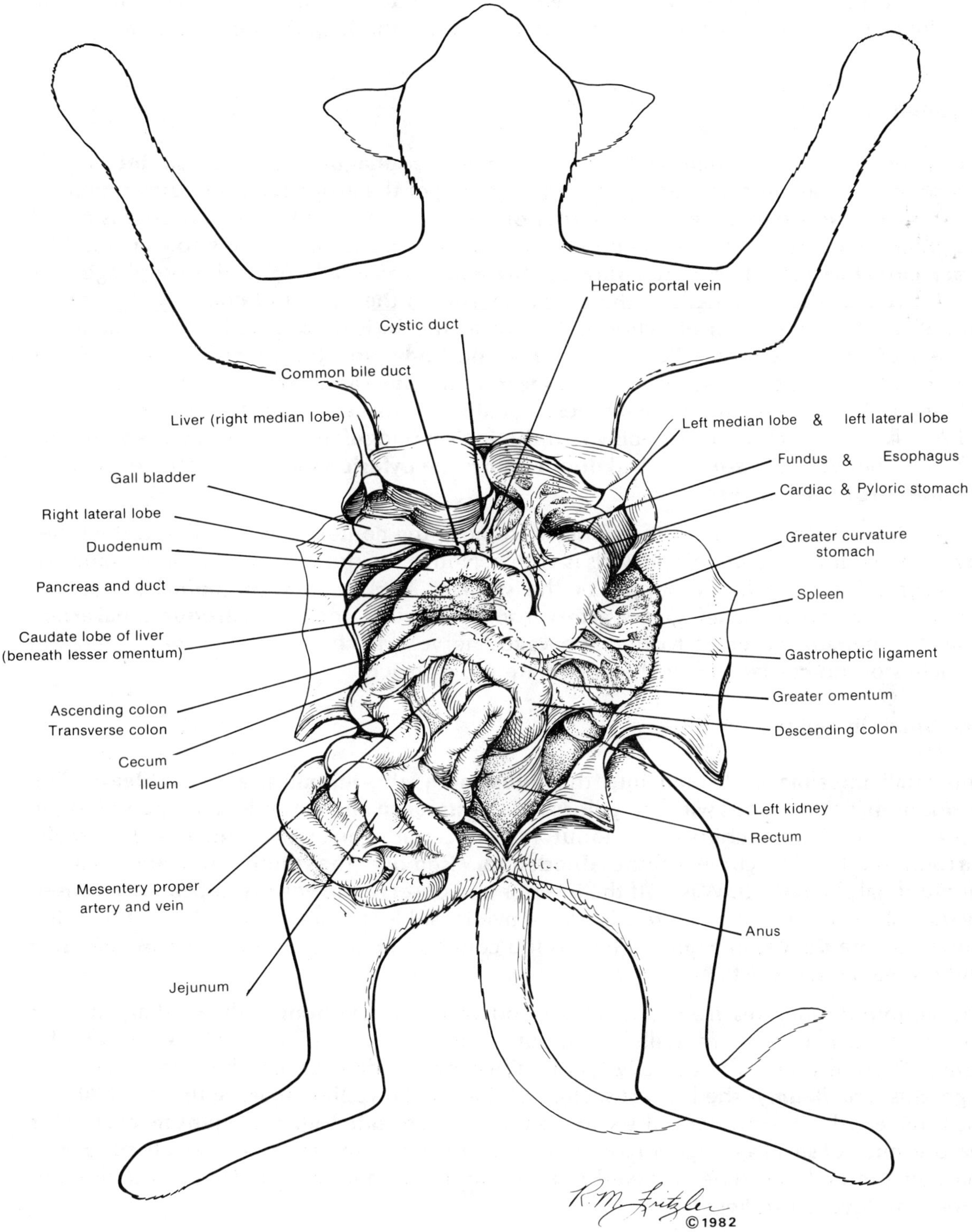

surface of the right median lobe of the liver. Identify and dissect the **cystic duct**, the **hepatic duct**, and the **common bile duct**. Trace the **common bile duct** to its entrance into the **duodenum**. Slit open the **duodenum** in this area, and identify the **duodenal papilla**, formerly called the **ampulla of Vater**. The pancreatic duct also enters the small intestine at this site. If you have a double-injected cat, dissect out the **hepatic portal system**.

Stomach

Locate the stomach, and identify the entrance of the **esophagus**, the **cardiac sphincter**. This ring of muscle can be palpated as a hard thickening of the esophagus at its attachment to the **stomach**. Notice that the lateral border of the stomach is convex. This border is called the **greater curvature**. The medial border is concave and shorter, leading to its name, the **lesser curvature**. The **lesser curvature** usually extends caudally from the **esophagus** for about two-thirds of its length. It then turns sharply to the right and cranially. The notch formed by this change in direction is known as the **incisura angularis**. The stomach is formed of three sections called the **fundus**, the **body**, and the **pylorus**. The **fundus** is bounded by the uppermost curve of the **greater curvature** as it extends above the **cardiac sphincter**. The **body** of the **stomach** lies caudally and makes up most of the organ. The **pylorus** is a constricted portion between the **incisura angularis** and the **duodenum**. The sphincter between the **stomach** and **duodenum** is the **pyloric sphincter**. It can be palpated like the **cardiac sphincter**.

Make a long slit in the wall of the **stomach**, remove the contents, and wash out the inner cavity so that the structure of the walls is visible. Notice the numerous folds of the **mucosa**, the **rugae.** Trace the blood supply of the stomach back to the **abdominal aorta.** The stomach is served by the **left gastric artery** from the **coeliac**, and the **gastroduodenal artery** from the **hepatic**. Attempt to find the vagus nerve branches to the stomach and the branches of the **coeliac plexus** that innervate the stomach.

The Small Intestine

The **small intestine** is divided into the **duodenum**, the **jejunum**, and the **ileum**. The **duodenum** is continuous with the **pylorus** of the **stomach** and extends caudally and to the right in a C-shaped course for about fifteen centimeters. The **pancreas** is partially surrounded by this curve of the **duodenum**. Remove the **duodenum** and make a longitudinal slit along it. Wash off the interior and observe the velvety appearance caused by the **villi**. About two centimeters from the pyloric sphincter, a small dilation in the wall is caused by the **duodenal papilla**, which is the common opening of the **common bile duct** and the **pancreatic duct**.

The **jejunum** continues from the caudal end of the **duodenum** with no sharp line of demarcation. It extends caudally for about twenty-five centimeters, where it joins the **ileum**. There is no microscopically visible difference to differentiate between these two segments. The **ileum** is the longest section of the **small intestine**, making up about 80% of the total length. The interior is the same as that of the **duodenum** and **jejunum**, except for the presence of small aggregations of lymphatic tissue that interrupt the velvety surface so that sometimes they can be observed without the aid of a microscope. These structures are known as **Peyer's patches**.

The Large Intestine

The **ileum** joins the **large intestine** in a blind pouch called the **cecum**. Usually this structure is surrounded by several large lymph nodes. In the human, the **vermiform appendix** protrudes from the base of this pouch. In the cat, however, there is no appendix.

From the **cecum**, the **large intestine** ascends along the right side of the abdominal cavity. At about the level of the third lumbar vertebrae, it turns left into the **hepatic flexure**. It continues left as the **transverse colon** until just under the **spleen**, where it turns down in the **splenic flexure** and descends into the **rectum**. The **colon** of the cat does not exhibit the clear-cut ascending, **transverse**, and **descending** sections that the human **colon** does. It is comparatively shorter than the human **colon**. Does this observation bear any relation to the diet of the cat compared to the human?

EXERCISE 3
MICROSCOPIC ANATOMY OF THE DIGESTIVE TRACT

The Esophagus

Study a cross section of mammalian esophagus, and observe the following structures.

1. **Serosa**

   An outer layer of loose connective tissue.

2. **Muscularis**

   The outer part consisting of longitudinally oriented muscle fibers and the inner part of circularly arranged fibers. In the upper quarter of the esophagus, these fibers are striated muscle; in the second quarter, bundles of smooth muscle appear among the striated; and in the lower third, the muscle is entirely smooth or viscera.

3. **Submucosa**

   A layer of connective tissue containing esophageal glands, which secrete mucous. A network of blood vessels also occurs in the submucosa.

4. **Mucosa**

   This consists of three layers: the **mucous membrane** of stratified squamous epithelium; the **lamina propria**, a thin layer of loose, connective tissue; and the **muscularis mucosa**, a thin layer of smooth muscle cells.

Draw a diagram of the layers of the esophagus, and in a "blow up" to one side, sketch in a few cells and fibers as observed in the different layers.

Figure 13-4. Lining of esophagus. E—stratified squamous epithelium; S—submucosa; V—blood vessels in submucosa.

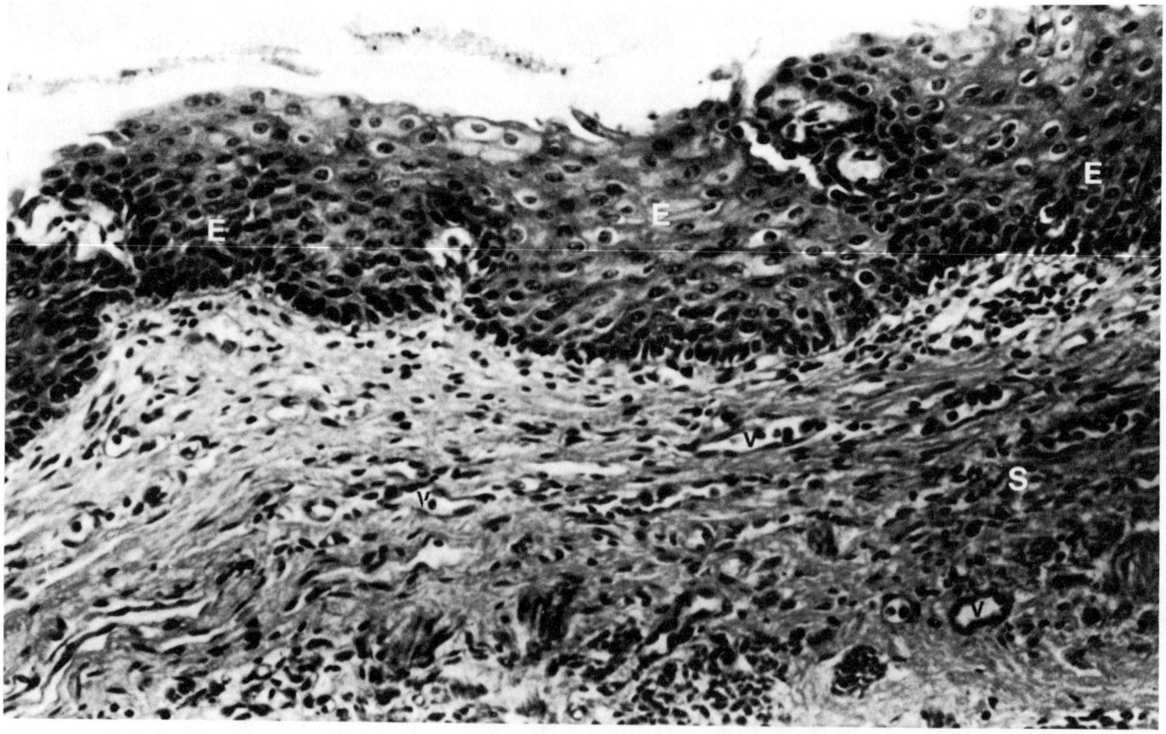

The Stomach

Study sections of the stomach and locate the structures listed below.

1. **Serosa**

   Connective tissue coating that is continuous with the large and small omenta.

2. **Muscularis**

3. **Submucosa**

4. **Mucosa**

   This layer is highly specialized throughout the digestive tract, according to the specialized functions of the different sections of the tract. In the stomach, the mucosa is composed of millions of glands densely arranged perpendicularly to the surface. There are three types of glands, different slightly in morphology and quite distinctly in function. The glands are usually named according to the region of the stomach in which they are predominately found. The **cardiac glands** are simple or compound tabular structures containing **mucous**-secreting cells along with a few **parietal cells**, which secrete HCl. The **fundic glands** are most numerous and possess the most complex morphology to match the larger number of secretory functions that they perform. They contain **mucous**-secreting cells, which are located generally in the neck of the gland, and **parietal cells**, as in the **cardiac glands**. Besides these two cell types, the lower portion of the glands contains **chief** or **zymogenic cells**, which secrete pepsin and gastric lipase. The **pyloric glands** are deeply penetrating coiled tubules that contain only one cell type, the **pale** staining, mucous-secreting variety.

Draw a typical **fundic gland**, showing the location of the different cell types.

What hormone is secreted by the gastric mucosa?

____

What is its function?

____

The Small Intestine

Observe a cross section of small intestine and locate the structures listed below.

1. **Serosa**
2. **Muscularis**
3. **Submucosa**
4. **Mucosa**

As in the esophagus and the stomach, the mucosa of the small intestine is highly specialized, in this case for two functions: (1) the absorption of nutrients broken down by mechanical and chemical action in the upper part of the digestive tract; and (2) the secretion of enzymes that complete the digestive process and hormones that aid in the control of secretory processes.

The accomplishment of the first function, absorption, is facilitated by several devices that serve to increase the surface area available for absorption. There are large folds of mucosa and submucosa that lie in a circular pattern one-half to two-thirds of the distance around the lining of the small intestine. These are the **plica circularis**. They are much more numerous in the duodenum, become scarce in the ileum, and are absent entirely in the jejunum.

In addition to these structures, countless fingerlike or leaflike projections arise from the entire surface of the mucosa. These are the **villi**, observed already in the gross examination of the small intestine. They are from 0.5 to 1.5 mm in height, and there may be from ten to forty per square millimeter. In between the **villi** are numerous deep invaginations of the mucosa. These are the **Crypts of Lieberkuhn**, which apparently secrete the various enzymes and hormones found in the intestinal juice. Most of the mucosa is covered with **microvilli**, small, fingerlike projections from the free surface of the cell. They are very uniform in size, being from 1 to 1.5 microns high and 0.1 microns in diameter. Interspersed among the absorptive cells are large, mucous-secreting cells with a large pale staining vacuole opening to the free surface. These are the **goblet cells**. In the **Crypts of Lieberkuhn**, two other varieties of cells may be observed. The large **paneth cells** at the base of the crypts contain large secretory granules and apparently secrete the various enzymes contained in the intestinal secretion. Another cell type is the **argentaffin cell**, so named because of its affinity for silver stains. These cells appear to secret a vasoconstrictor.

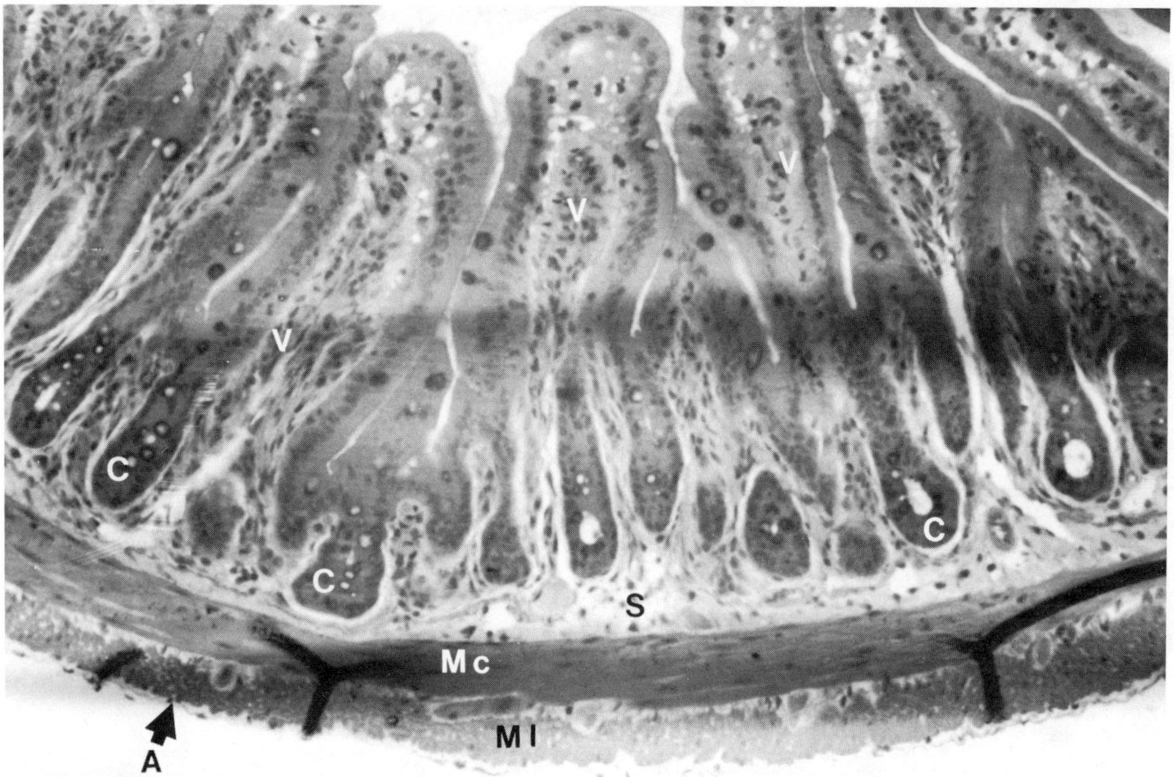

Figure 13-5. Low-power photomicrograph of mouse intestine. V—villi; C—Crypts of Lieberkuhn; S—submucosa; Mc—circular smooth muscle; Ml—longitudinal smooth muscle; A—thin layer of connective tissue, the adventitia.

Draw a diagram of the cross section of the small intestine, labeling the various layers of the plica circularis and the villi.

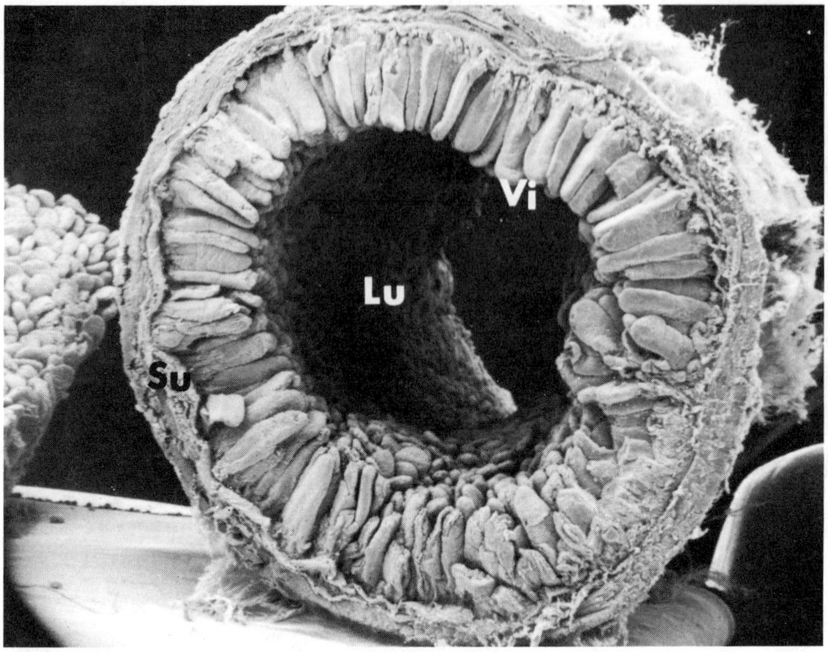

Figure 13-6. Scanning electron micrograph of a cross section of the small intestine. Vi—villi; Lu—lumen; Su— submucosa.

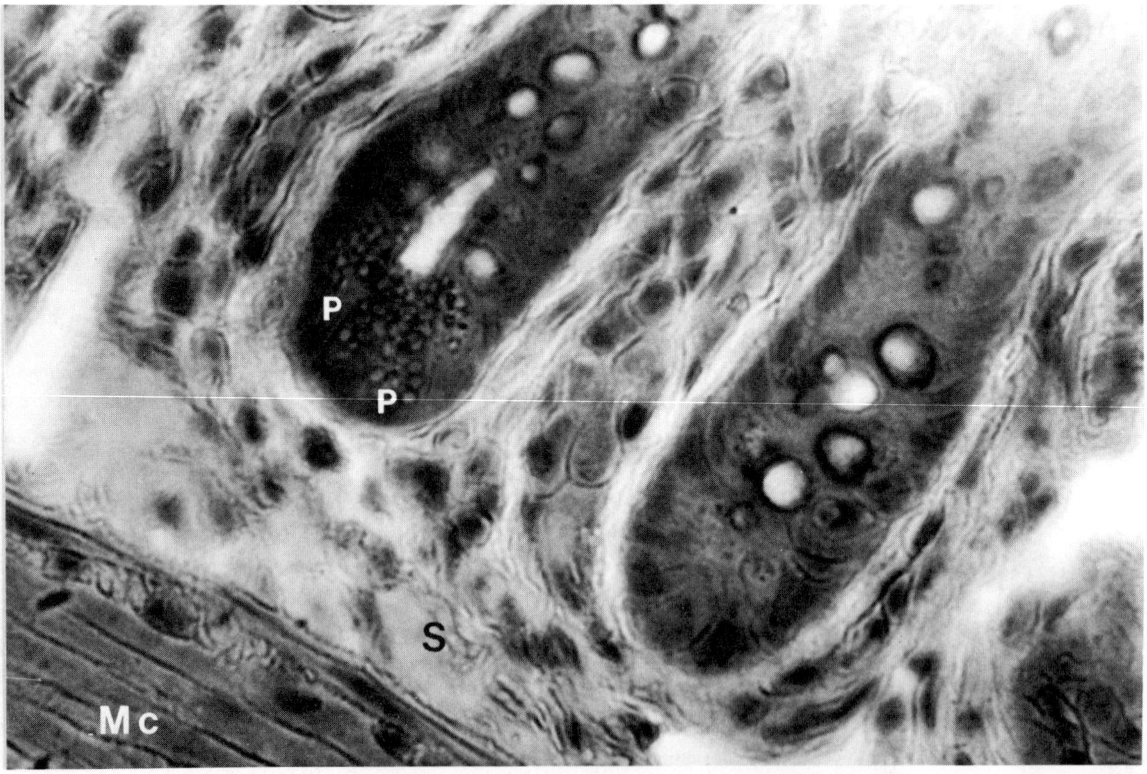

Figure 13-7. High-power photomicrograph showing two Crypts of Lieberkuhn, with paneth cells (P) filled with secretory vesicles. S—submucosa; Mc—circular layer of smooth muscle. Locate this area on Figure 13-5 and outline it.

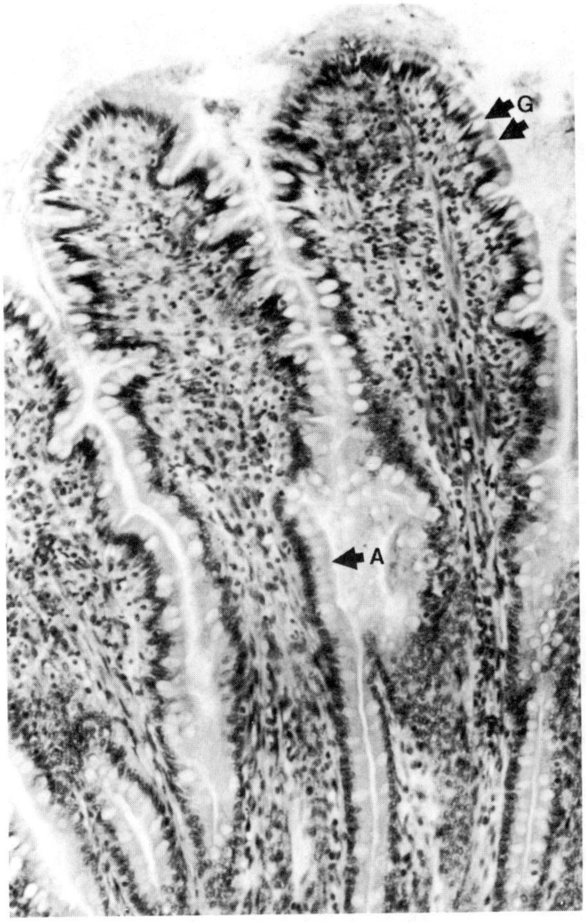

Figure 13-8. Villi of small intestine. A—simple columnar epithelial absorptive cells; G—goblet cells, unicellular glands secreting mucous.

Draw a pair of villi and an associated **Crypt of Lieberkuhn**, labeling all cell types that can be observed.

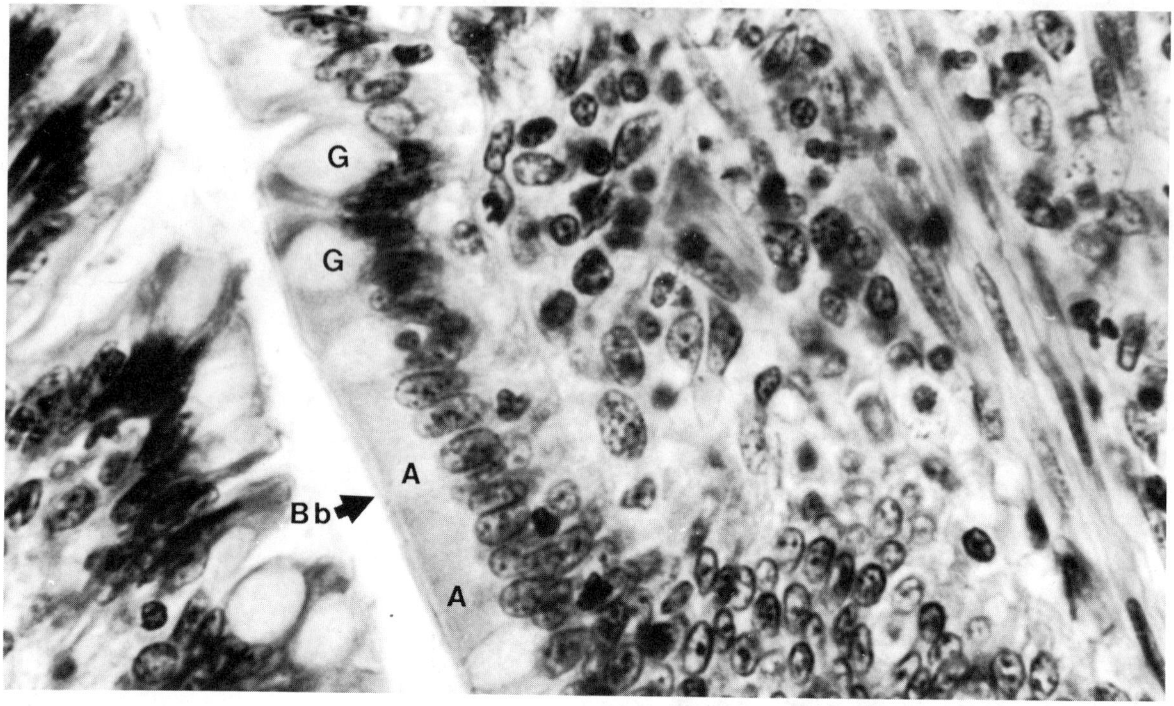

Figure 13-9. High-power photomicrograph of a portion of a villus showing: Bb—brush border (microvilli) of A—absorptive cells and G—goblet cells.

The Liver

The **liver lobule** is the key to understanding the microscopic structure of the liver. The best procedure is to locate a **central vein** under low power, then look for a roughly hexagonal pattern of radiating cell cords surrounding the **central vein**. At some of the corners of the hexagon, **interlobular veins** will be observed. These are the terminal branches of the **portal vein**, and they feed into the **sinusoidal blood spaces** between the cells of the liver lobule. In close association with the **interlobular veins**, small **bile ducts** and small branches of the **hepatic artery** may usually be seen. These three structures—the **interlobular vein**, the **bile duct branch**, and the **branch of the hepatic artery**—form a "**portal triad**."

The cells of the liver that are seen as cords radiating out from the **central canal** actually form walls for the irregular blood spaces lying between them. These blood spaces, or **sinusoids**, resemble capillaries in that they are lined with endothelium. However, in the **sinusoids**, the lining is in closer contact with the surrounding liver cells. Large phagocytic cells inhabit some areas of the sinusoid walls. These **Kupffer cells** ingest foreign material, worn-out red blood cells, and particulate matter.

1. Diagram a liver lobule, labeling the structures listed above.

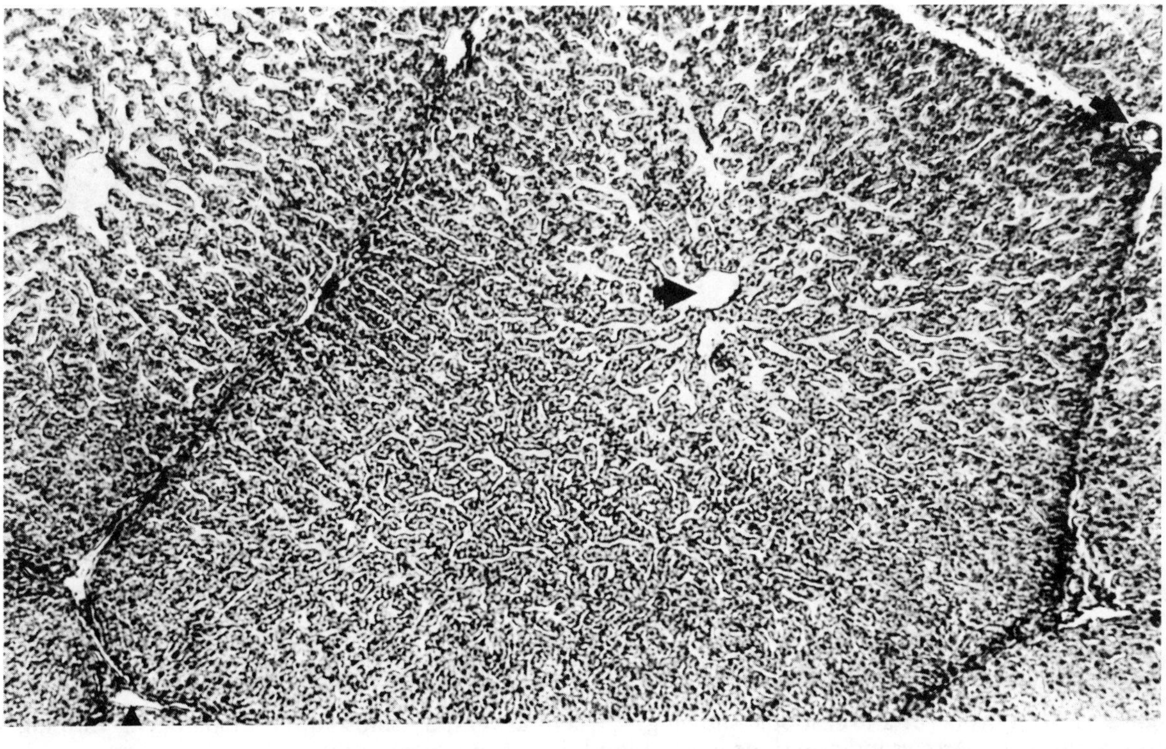

Figure 13-10. Low-power photomicrograph of a pig liver lobule. Center arrow points to a central vein, arrows at upper right and lower left indicate portal triads.

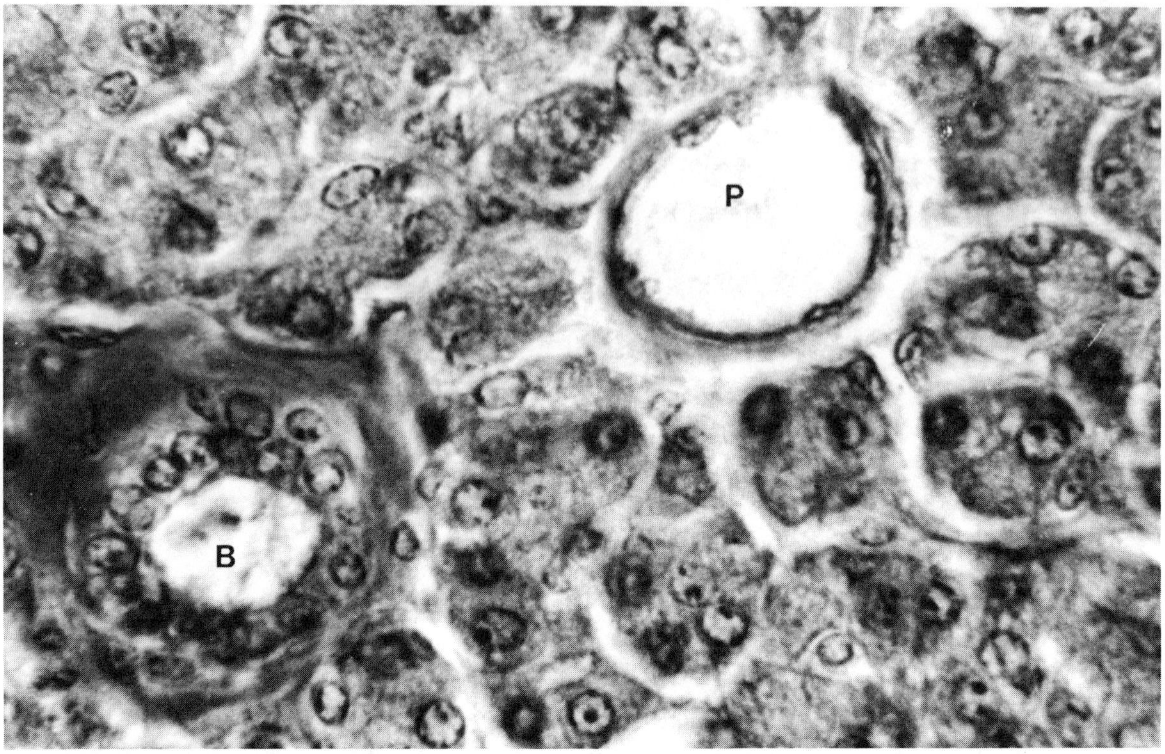

Figure 13-11. High-power photo showing two of the parts of a portal triad, a branch of the portal vein (P), and a branch of the bile duct (B).

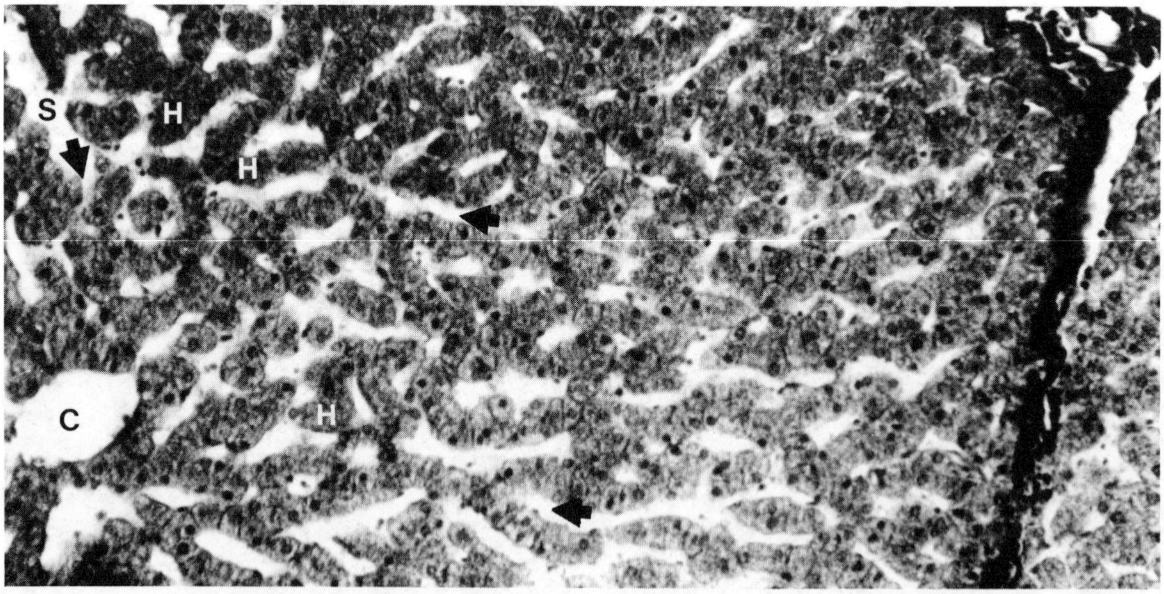

Figure 13-12. Portion of a liver lobule showing: C—central vein; S—sinusoid; H—hepatocytes forming walls of sinusoids. Arrows indicate direction of blood flow through sinusoids.

The Pancreas

The pancreas is a large gland with both an exocrine and an endocrine function. The exocrine portion secretes several digestive enzymes, and the endocrine cells secrete insulin and glucagon, two hormones concerned with the regulation of metabolism.

Exocrine Portion

This portion consists of a compound acinus gland, the lobules of which are held together by connective tissue. Locate under low power a group of cells clustered around a central duct. Observe the shape of the cells, the location of the nucleus, and the numerous secretion granules present.

Sketch one of these clumps (more properly called an **acinus**). Label the features mentioned above.

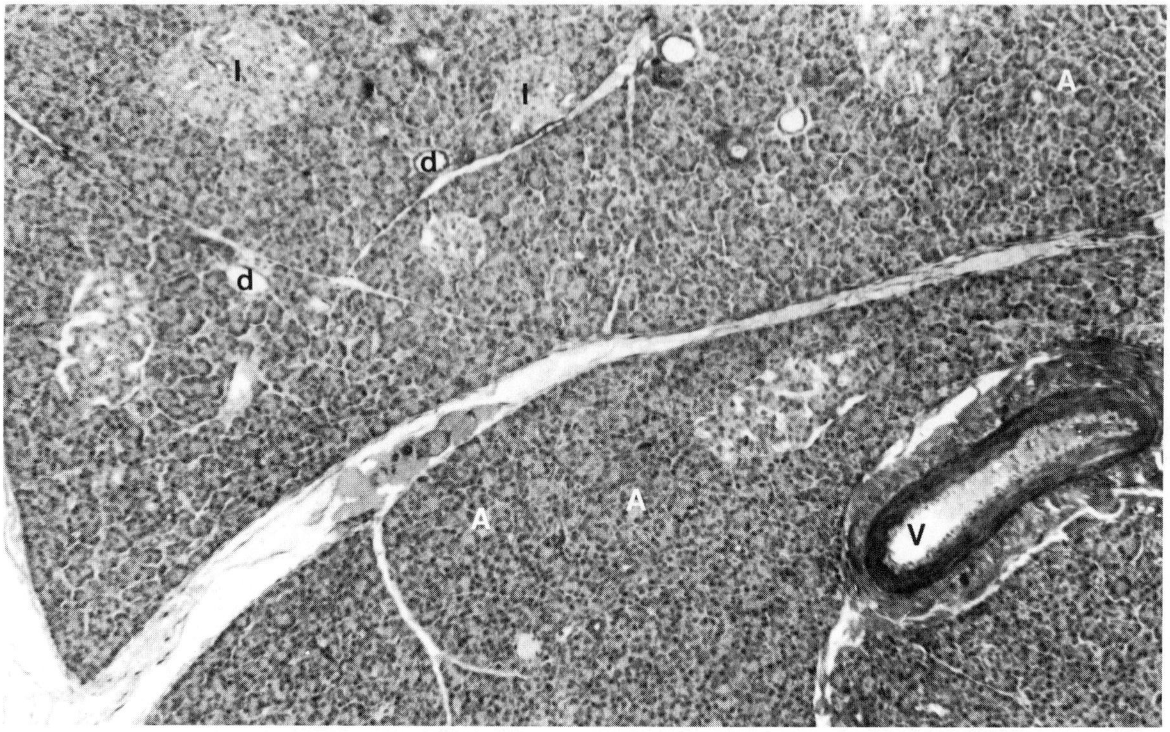

Figure 13-13. Low-power photomicrograph of a section of pancreas. A—clumps of acinar cells forming the exocrine glands; I—Islets of Langerhans, the endocrine glands of the pancreas; d—branches of ducts of the exocrine glands; V—large vein.

LABORATORY 14

Structure of the Urinary and Reproductive Systems

OBJECTIVES

Following this laboratory session, you should be able to accomplish the following objectives.

Gross Anatomy of the Urinary System

Identify the following structures on photographs, drawings, models, or a cat dissection.

1. Capsule
2. Lateral border
3. Medial border
4. Renal artery
5. Renal vein
6. Cortex
7. Medulla
8. Ureter
9. Minor calyces
10. Major calyx
11. Renal papilla
12. Renal columns
13. Pyramids
14. Intralobular veins and arteries
15. Bladder
16. Urethra

Microscopic Anatomy

Identify the following structures on microscope slides, photomicrographs, or drawings.

1. Bowman's capsule
2. Glomerulus
3. Afferent arteriole
4. Efferent arteriole
5. Peritubular capillaries (vasa recta)
6. Distal convoluted tubule
7. Proximal convoluted tubule
8. Loop of Henle
9. Collecting tubule
10. Juxtaglomerular apparatus

Gross Anatomy of the Reproductive Tract

Identify the following structures on photographs, drawings, or a cat dissection.

1. Female — ovaries, uterine horns
2. Male — testes, epididymis, vas deferens

Histology of the Testes and Ovaries

Identify the following structures on photographs, drawings, or microscope slides:

Testes

1. Tunica albuginea
2. Seminiferous tubules
3. Interstitial cells (of Leydig)
4. Spermatogonia
5. Spermatocytes
6. Spermatids
7. Spermatozoa

Ovaries

1. Germinal epithelium
2. Primary follicle
3. Oocyte
4. Follicular cells
5. Maturing follicles
6. Antrum
7. Theca folliculi
8. Corpus luteum

EXERCISE 1
ANATOMY OF THE URINARY SYSTEM

Dissection of the Urinary System

The Kidney

1. Locate the **kidney**. Where does it lie with respect to the parietal peritoneum?

   ____

2. Trace the route of the **renal artery** and **vein** into the **hilum**.

3. Identify and trace the course of the **ureter** as it emerges from the **hilum**, across the **psoas** muscle and into the lateral part of the posterior surface of the **urinary bladder**.

4. Remove one **kidney** from the cat, trying to maintain its relationship to the **adrenal gland**. Make a midsagittal section through the **kidney**, and examine the interior.

5. Pull a small section of the connective tissue **capsule** from the surface. Observe the concave medial border and the convex lateral border. Notice the distinct color difference between the outer portion, the **cortex**, and the inner section, or **medulla**. How are these differences related to the microscopic structure of the **kidney**?

____

____

6. Study the relationships between the **renal columns**, **renal papillae**, **pyramids**, and **major** and **minor calyces**.

7. Label Figure 14-1 with the following structures.

Capsule	Ureter
Lateral border	Minor calyx
Medial border	Major calyx
Renal artery	Renal papilla
Renal vein	Renal column
Cortex	Pyramid
Medulla	

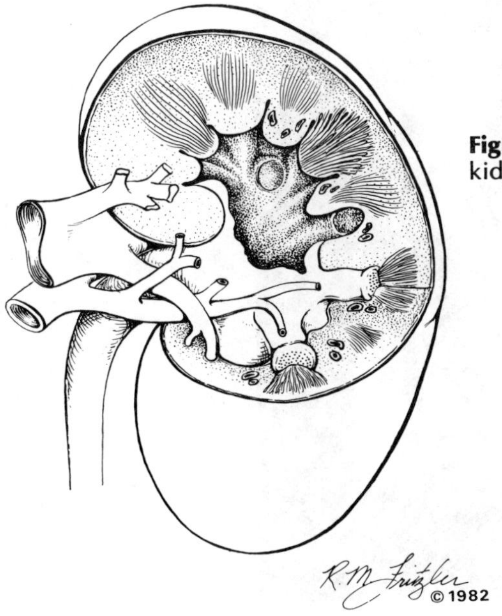

Figure 14-1. Longitudinal section of the human kidney.

The Bladder

1. Trace the **ureters** to their entrance into the **bladder.**
2. Observe the location, attachments, and, if possible, the innervation of the **bladder** (the **inferior hemorrhoidal nerve**).
3. Slit the ventral wall of the **bladder**, and locate the entrance of the **ureters**. Use a probe if necessary.
4. Locate the exit of the **urethra**, and probe this tube. Can you identify the **urethral sphincters**?

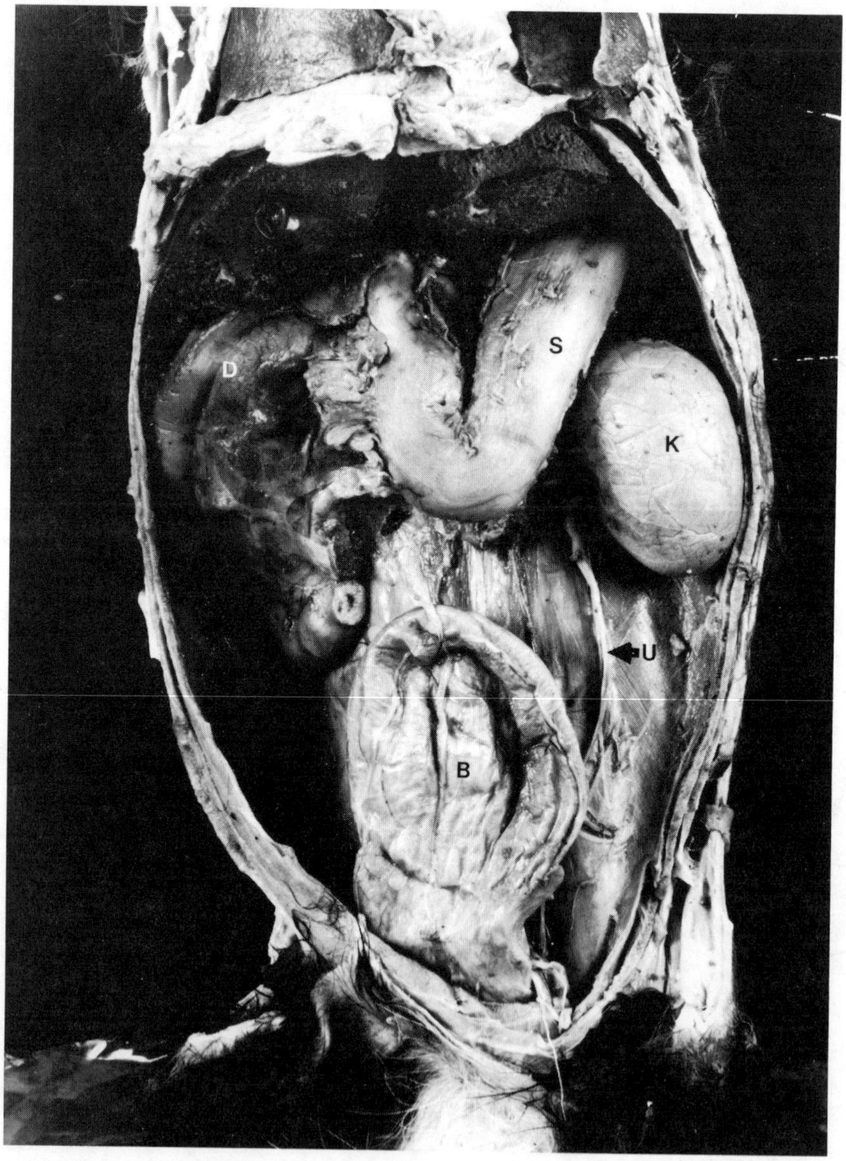

Figure 14-2. Abdominal cavity with intestine removed. S—stomach; U—ureter; K—kidney; D—duodenum; B—bladder.

EXERCISE 2
HISTOLOGY OF THE KIDNEY

Locate the following structures on microscope slides, and label them on Figure 14-3.

1. Glomerulus
2. Proximal convoluted tubule
3. Distal convoluted tubule
4. Descending limb of Loop of Henle
5. Ascending limb of Loop of Henle
6. Collecting tubule
7. Bowman's capsule

Figure 14-3.

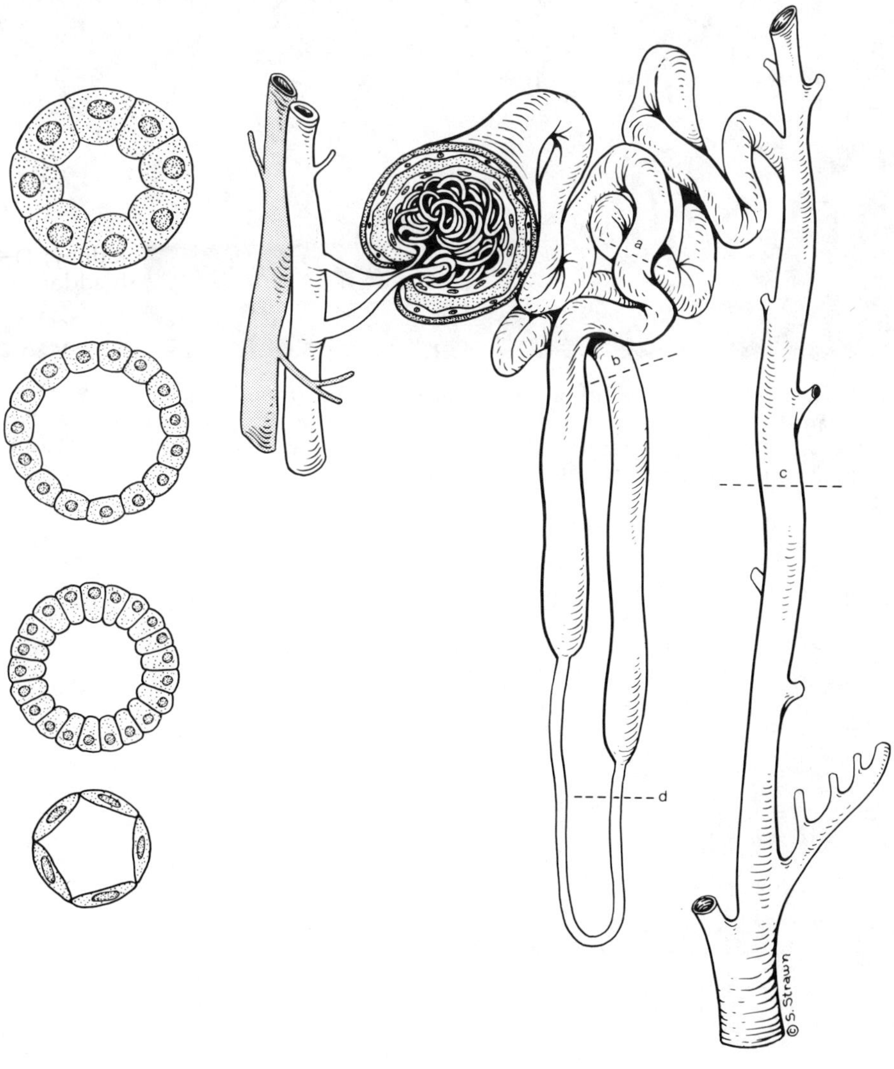

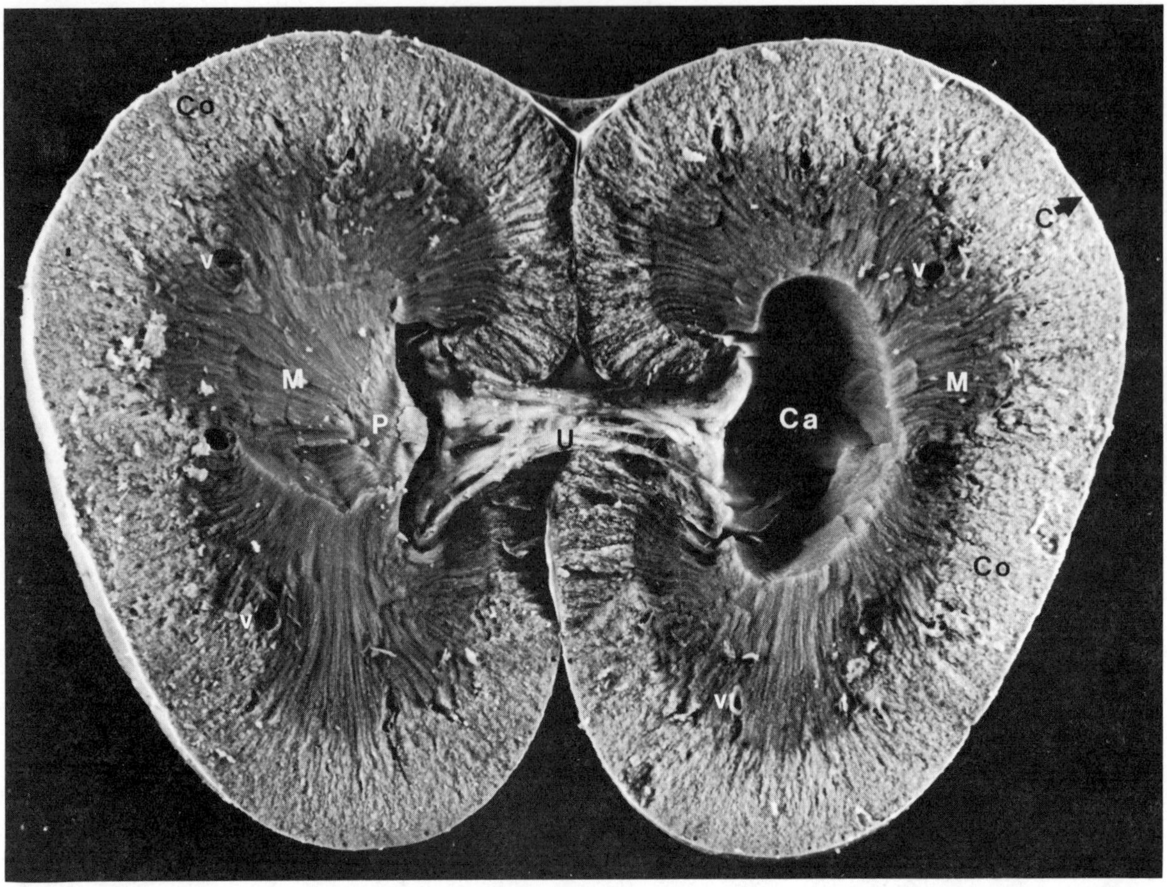

Figure 14-4. Enlarged view of a cat left kidney sectioned longitudinally. C—capsule; Co—cortex (lighter granular area); M—medulla (darker striated area; P—pyramid (the cat has a unilobar kidney; Ca—calyx with part of pyramid removed; U—ureter split longitudinally. Notice numerous blood vessels to the right of the small letter v.

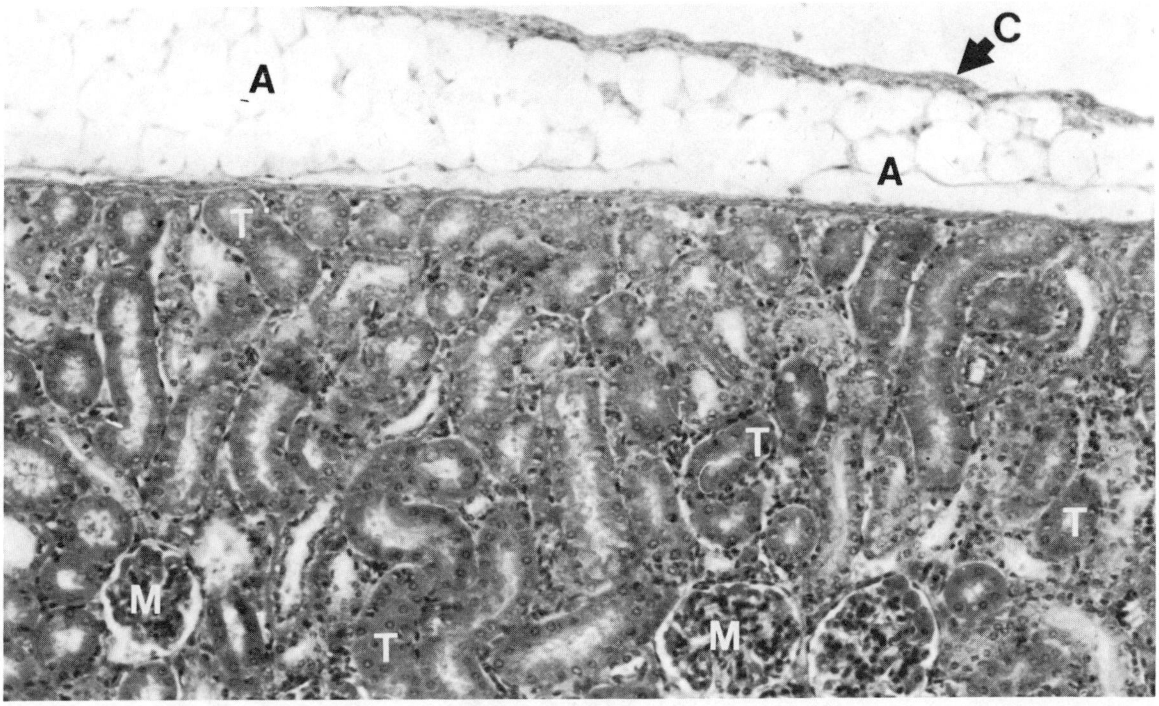

Figure 14-5. Kidney cortex, 50X. C—capsule; T—tubule; M—malpighian body; A—underlying adipose tissue.

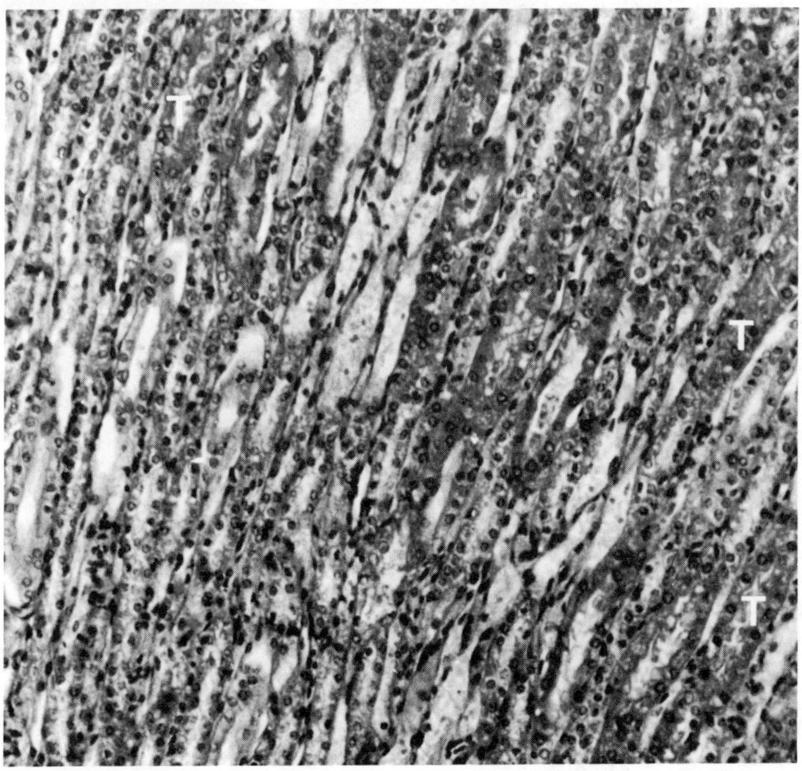

Figure 14-6. Kidney medulla, 50X. T—tubules oriented longitudinally. Capillaries are interspersed around the tubules.

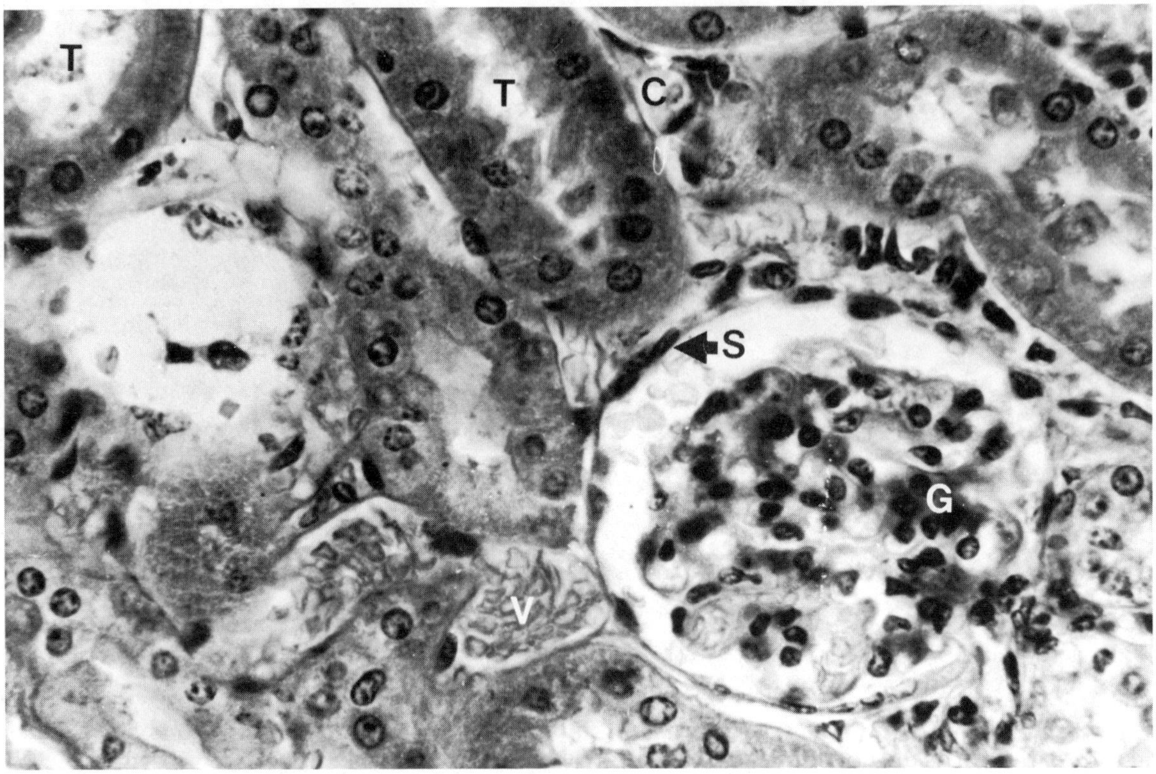

Figure 14-7. High-power photomicrograph of kidney cortex. T—tubules; C—capillary; V—blood vessel; S—squamous cells of Bowman's capsule; G—glomerulus.

EXERCISE 3
GROSS ANATOMY OF THE REPRODUCTIVE TRACT

Female Reproductive Organs

Locate the following structures on Figure 14-8.

1. **Ovaries**

   Remove the kidneys from a female cat, and obsere the two small "bean-shaped" structures located posteriorly to the kidneys. Remove them, slit them longitudinally, and observe them with a dissecting microscope. You will be able to see many of the parts discussed in the exercise on histology.

2. **Uterine horns**

   In the cat, as in most mammals, the **uterus** is composed of two long, tubular structures that originate at the **ovary** and fuse together just dorsal to the **urinary bladder** and **urethra**. These **uterine horns** are an adaptation to provide space for the development of several fetuses at one time, i.e., a "litter." If a pregnant cat is available, open the uterine horns and observe the size and degree of development of the fetuses. Normal gestation time for cats is about sixty days. Can you judge the age of the fetuses from their size compared to a newborn kitten?

Male Reproductive Tract

Locate the following structures on Figure 14-9.

1. **Testes**

   Locate the **scrotal sac**, just ventral to the **anus**, and carefully cut it open to expose the **testes**. Remove one, but leave the other in place so that later you can trace the **vas deferens** and **spermatic cord**. Cut the removed **testis** in half with a sharp scalpel or razor blade. Observe the internal structure under the dissecting microscope. This will help you understand the next section on histology.

2. **Epididymis**

   The **Epididymis** is the extension of the **seminiferous tubules** and forms a prominent bulge on the surface of the **testis** at the point of origin of the **vas deferens**.

3. **Vas deferens**

   This is the tube that conveys the sperm from the **testes**. As it passes from the scrotal sac to the abdominal cavity, the **vas deferens** is wrapped in a tough connective tissue sheath along with blood vessels and nerves. This is the **spermatic cord** and is supported by the **inguinal ligament** as it passes through the **inguinal canal**.

Figure 14-8. Female urogenital system of the cat.

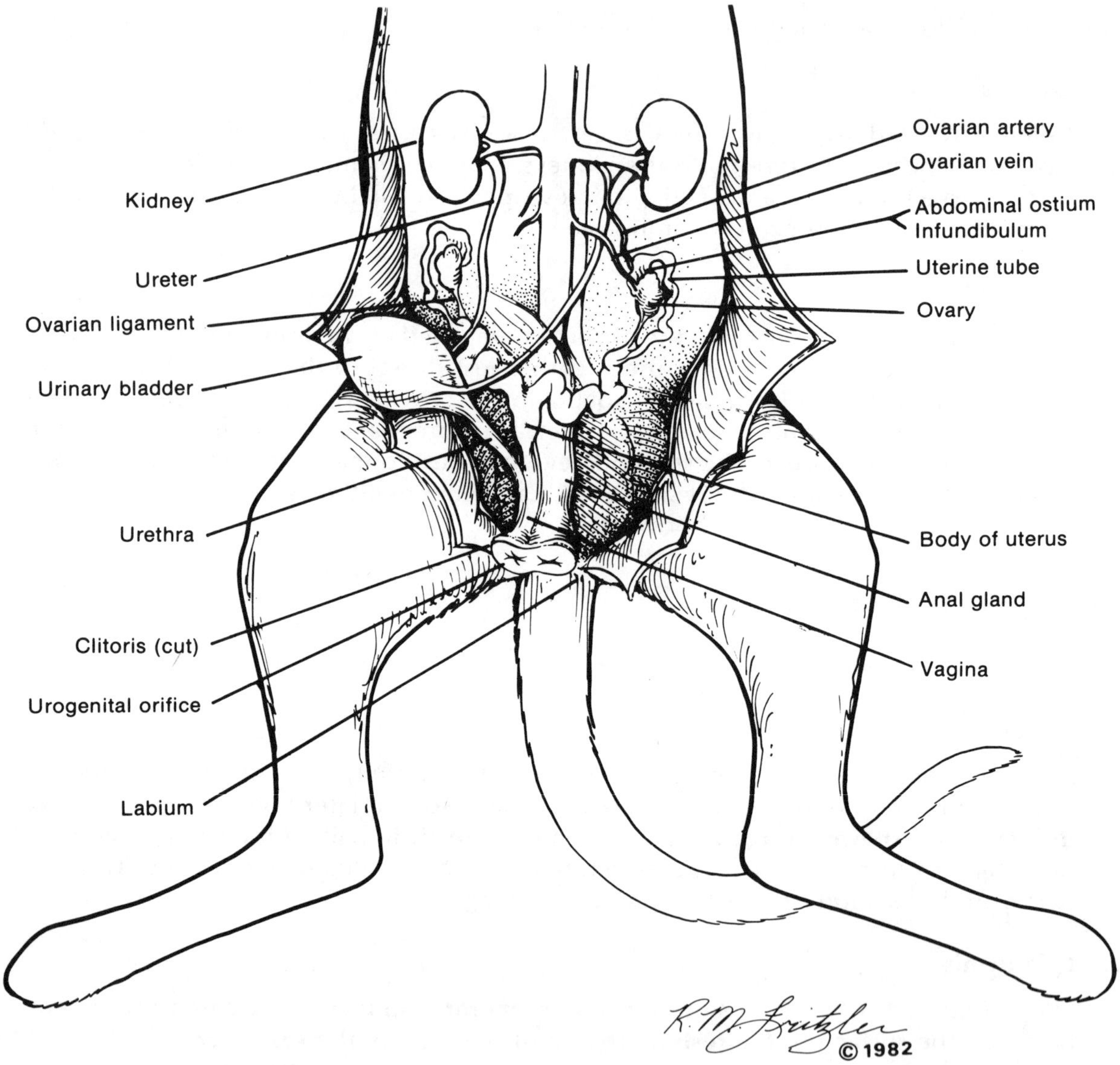

Figure 14-9. Male urogenital system of the cat.

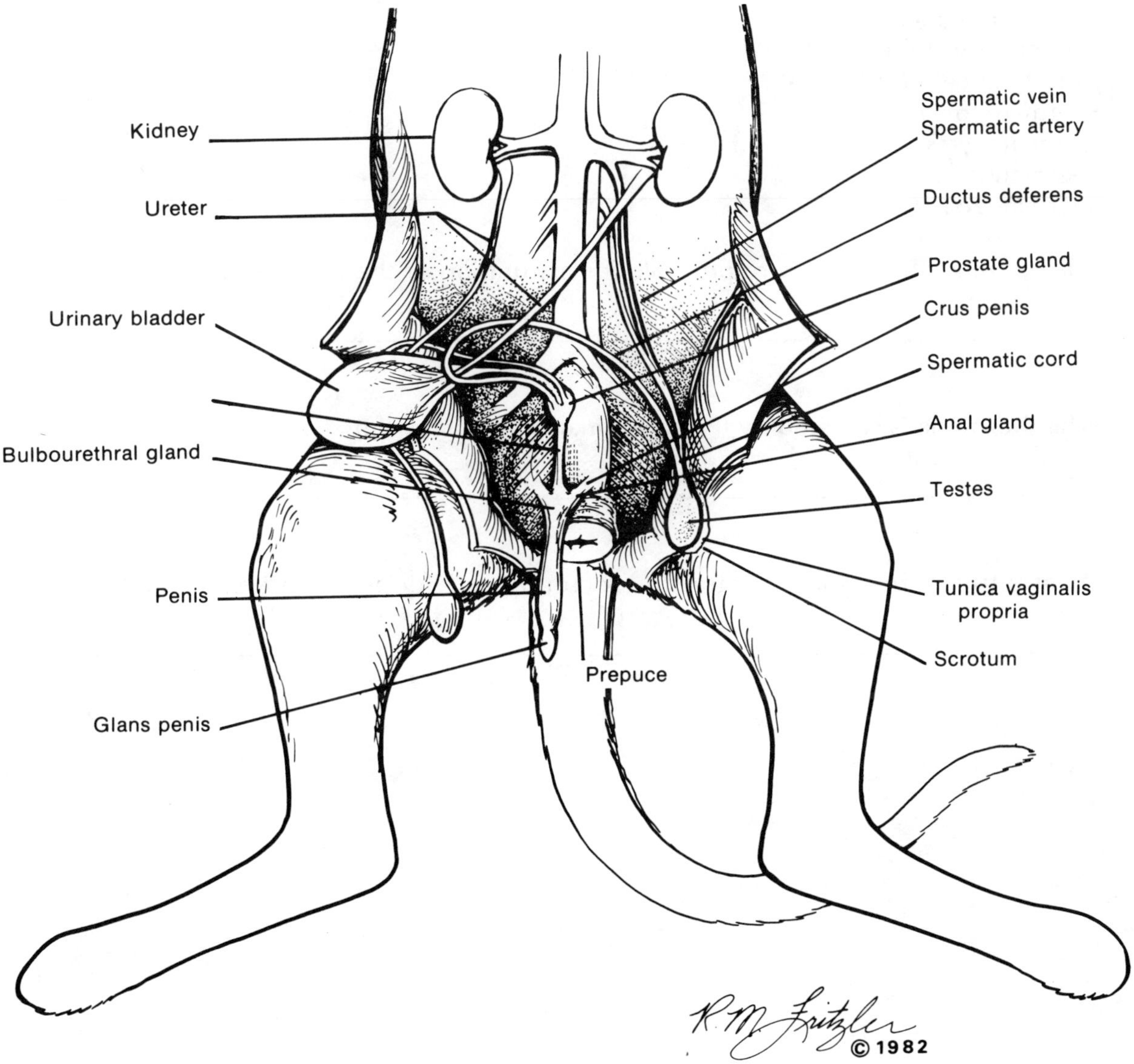

EXERCISE 4
HISTOLOGY OF THE TESTES

1. With the microscope, examine a prepared slide of rat testes under low and high power. Locate the following structures, and sketch them in the space below.

   Tunica albuginea
   Seminiferous tubules
   Interstitial cells (of Leydig)
   Spermatogonia
   Spermatocytes
   Spermatids
   Spermatozoa

2. What pituitary hormone is responsible for maturation of the spermatozoa?

   ____

3. What hormone is secreted by the interstitial cells (of Leydig)?

   ____

4. What pituitary hormone initiates the secretory activity of the interstitial cells?

   ____

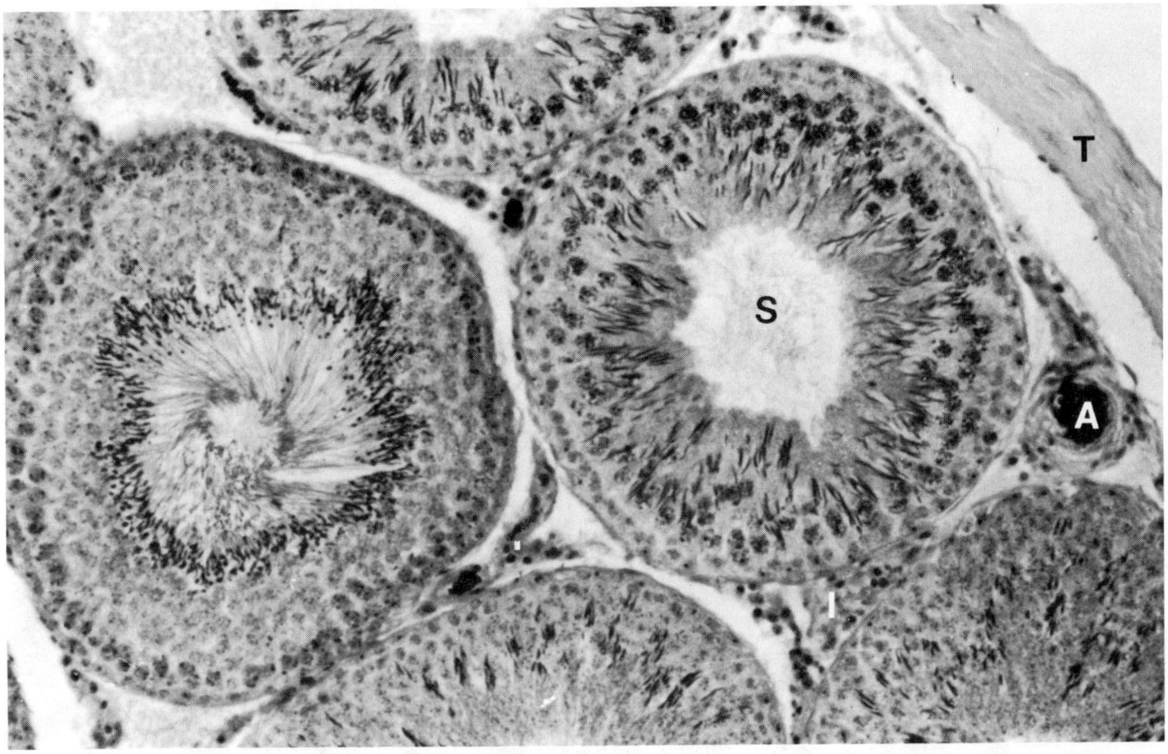

Figure 14-10. Low-power photomicrograph of a section of rat testes showing a complete seminiferous tubule and parts of three others. T—tunica albuginea, a dense fibrous connective tissue of the testis; S—lumen of seminiferous tubule; I—interstitial cells (of Leydig); A—small artery.

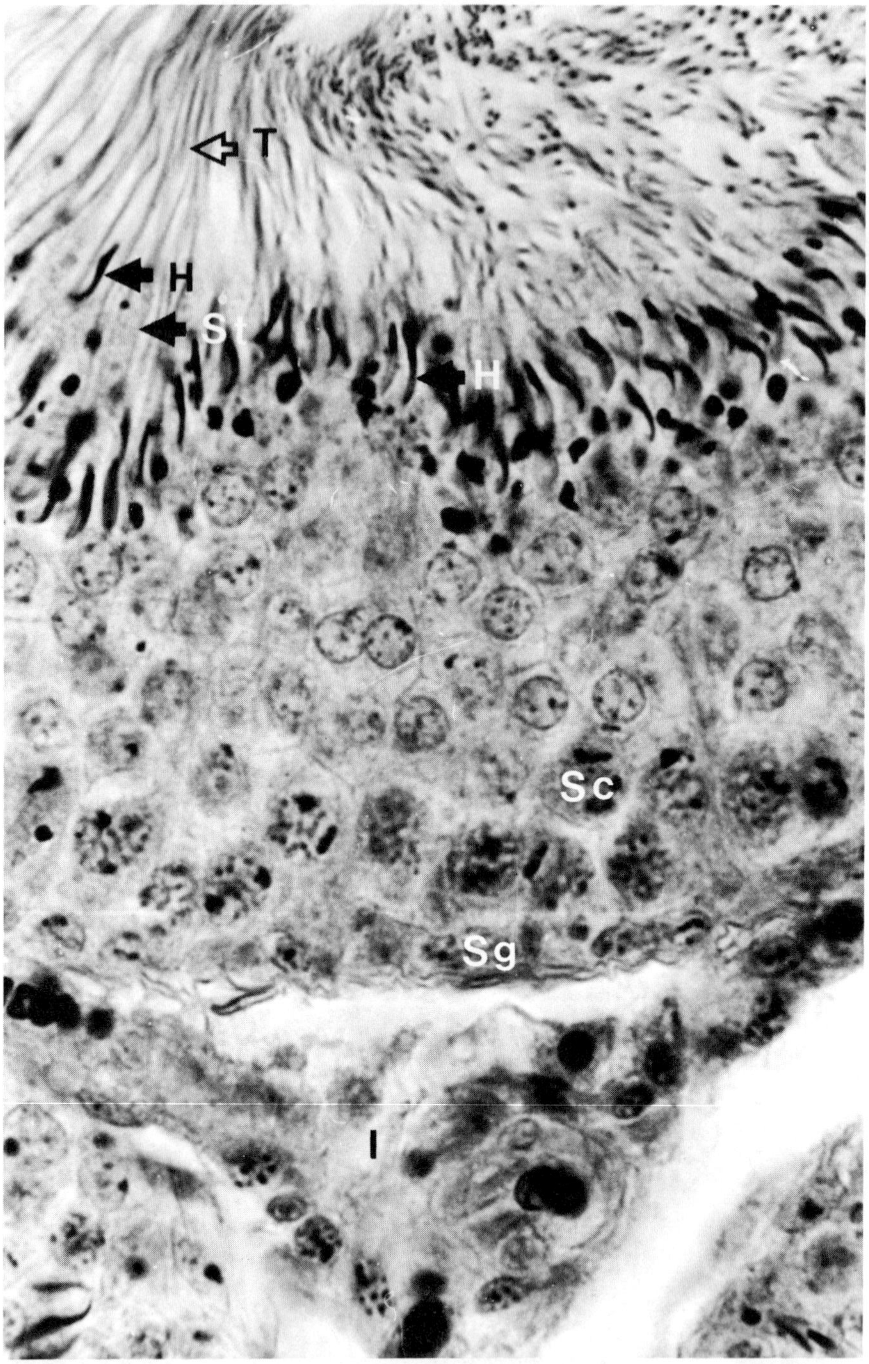

Figure 14-11. High-power photomicrograph of a portion of a seminiferous tubule. I—interstitial cells (of Leydig); Sg—spermatogonium; Sc—spermatocyte; St—spermatid (late); H—head of spermatozoan protruding into lumen.

EXERCISE 5
HISTOLOGY OF THE OVARY

Examine a slide of ovary with the aid of a microscope. Locate the following structures. Sketch them in the space provided below.

1. Germinal epithelium
2. Primary follicle
3. Oocyte
4. Follicular cells
5. Maturing follicles
6. Antrum
7. Theca folliculi
8. Corpus luteum

1. What pituitary hormones are responsible for the maturation of the ovum and ovulation?

____

2. What hormone is secreted by the ovarian follicle?

____

3. What is its function?

____

____

4. What hormone is secreted by the corpus luteum?

____

5. What is its function?

____

____

6. What pituitary hormone is responsible for the conversion of a ruptured follicle into a corpus luteum?

____

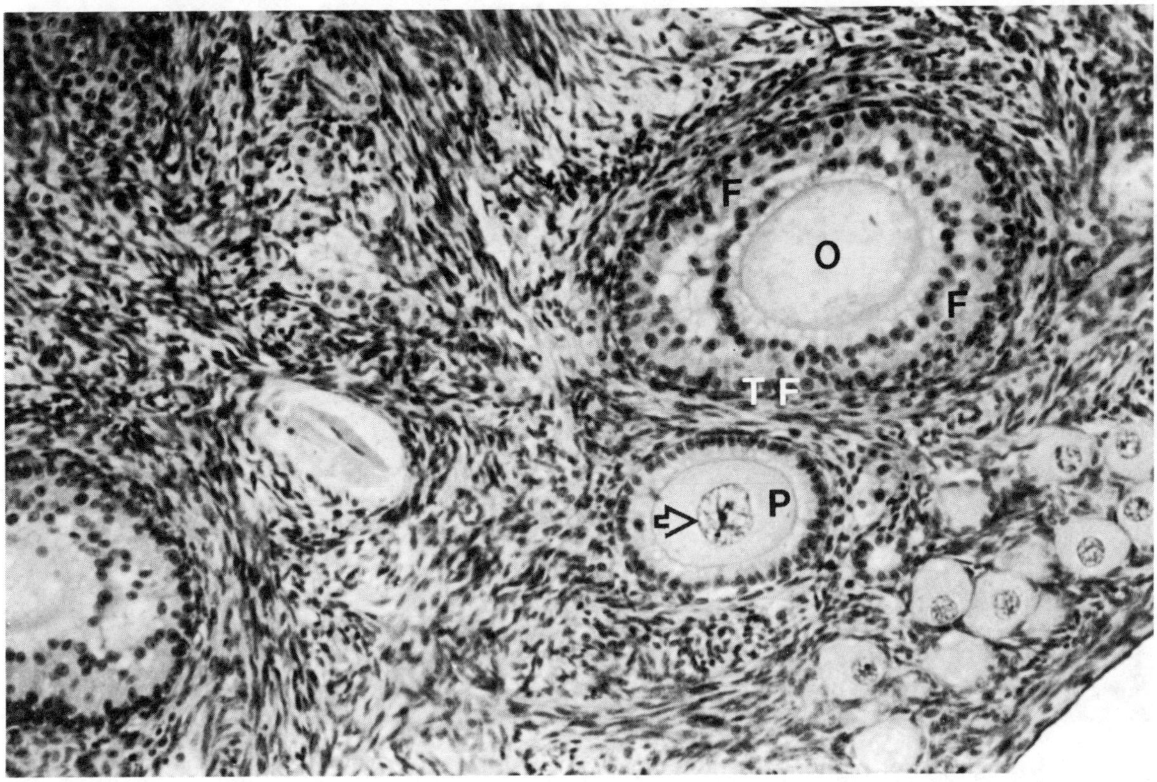

Figure 14-12. Low-power photomicrograph of a section of cat ovary showing an egg "nest" at the lower right. P—oocyte in a primary follicle, arrow points to nucleus; O—oocyte in a maturing follicle, nucleus not in plane of section; F—follicular cells of maturing follicles; Tf—theca foliculi.

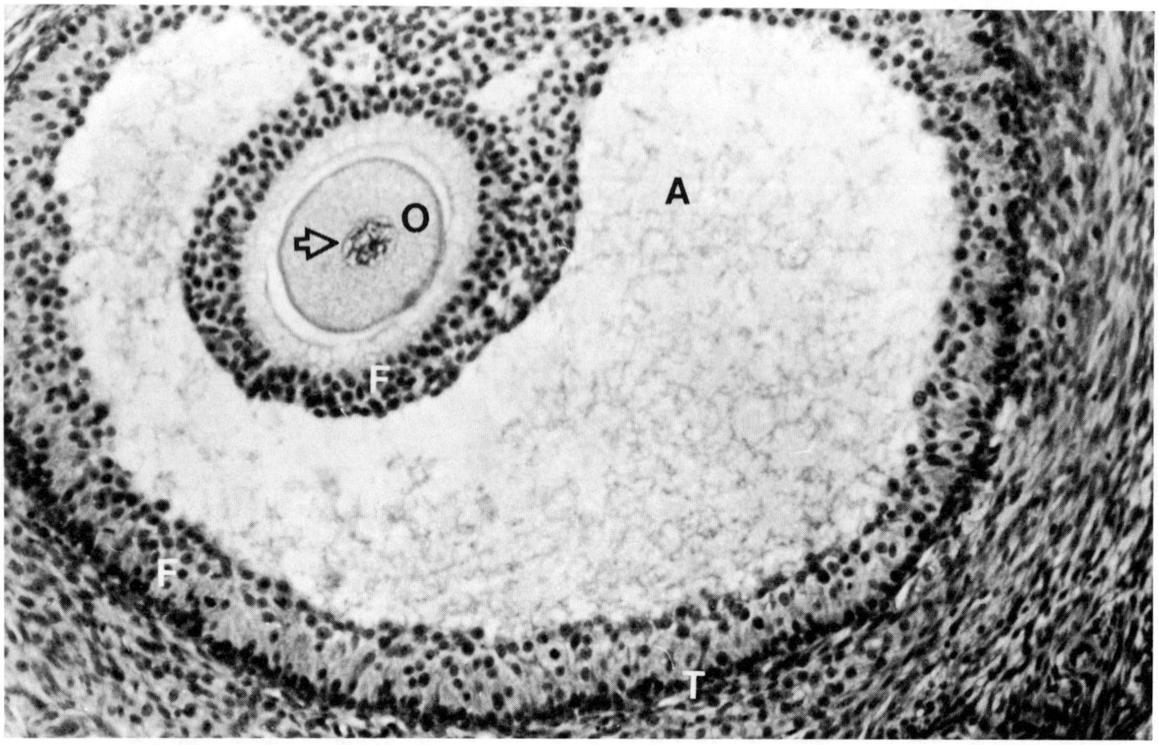

Figure 14-13. High-power photomicrograph of a nearly mature follicle. T—theca folliculi; F—follicular cells, some form several layers lining the follicle and others form several layers surrounding the oocyte; A—antrum, fluid-filled space in follicle; O—oocyte, arrow points to nucleus.

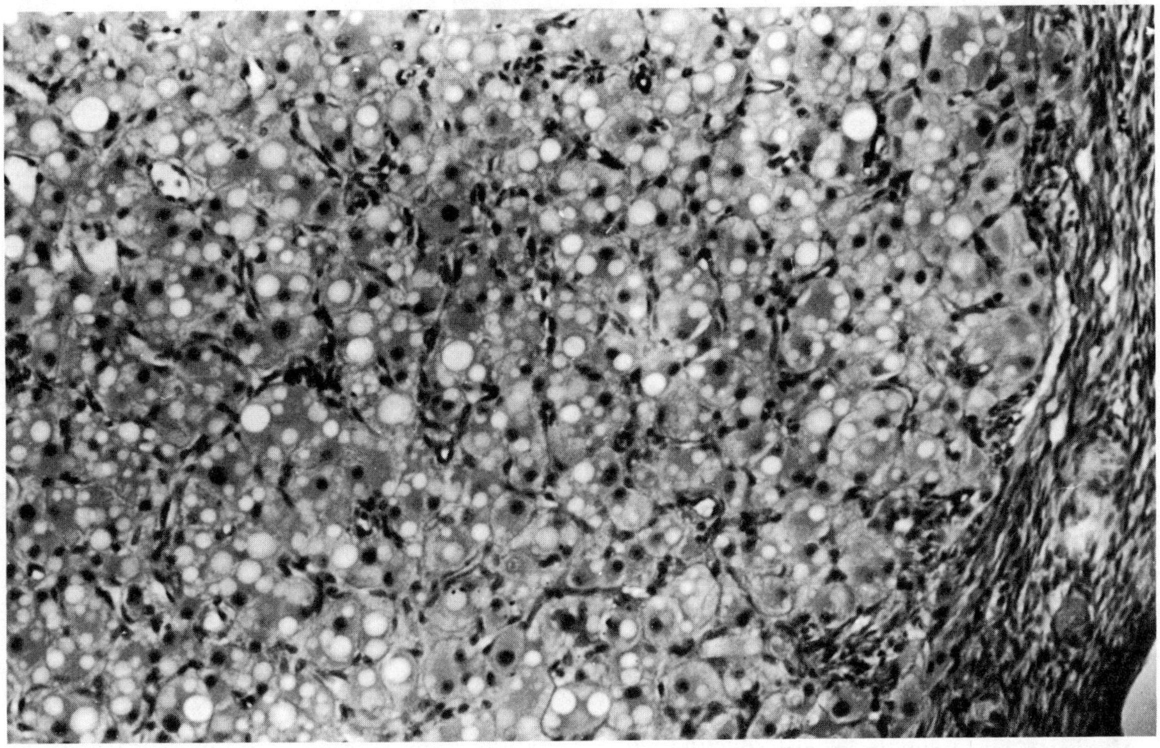

Figure 14-14. High-power photomicrograph of a portion of the cells of the corpus luteum. Note the large lipid vacuoles, probably progesterone or a precursor.

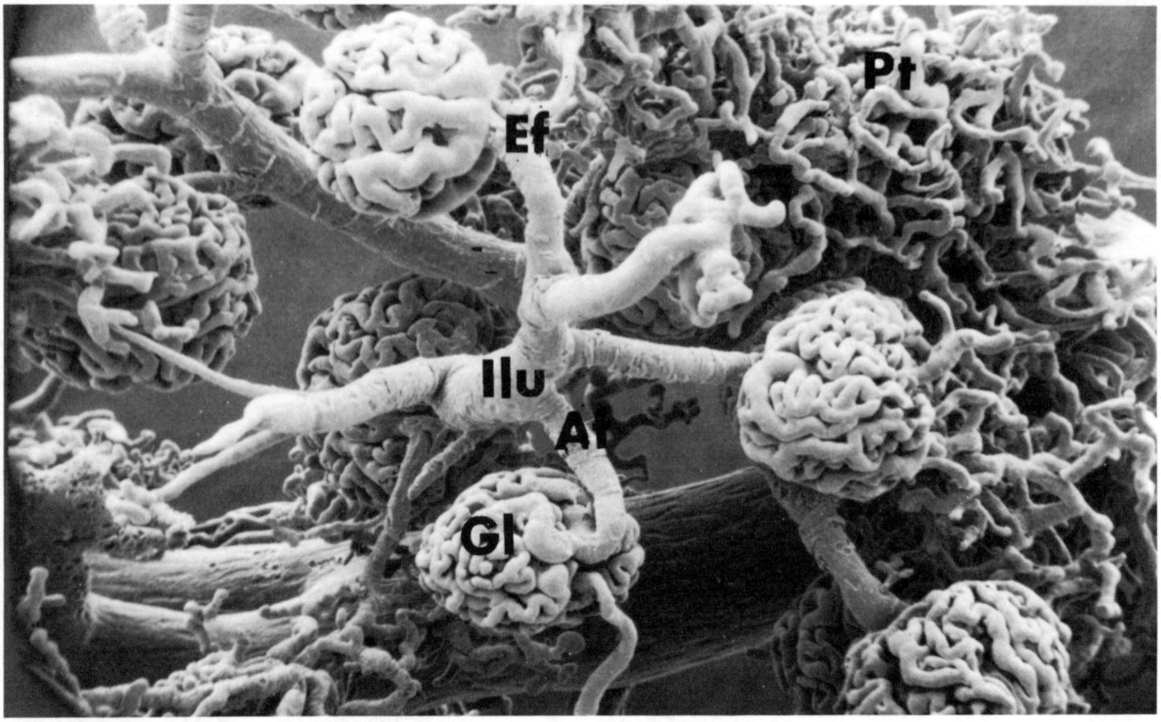

Figure 14-15. Scanning electron micrograph of a portion of the circulation of the cortex, 260X. Gl—glomerulus; Af—afferent arterioles; Pt—peritubular capillaries; Ef—efferent arterioles; Ilu—interlobular artery.

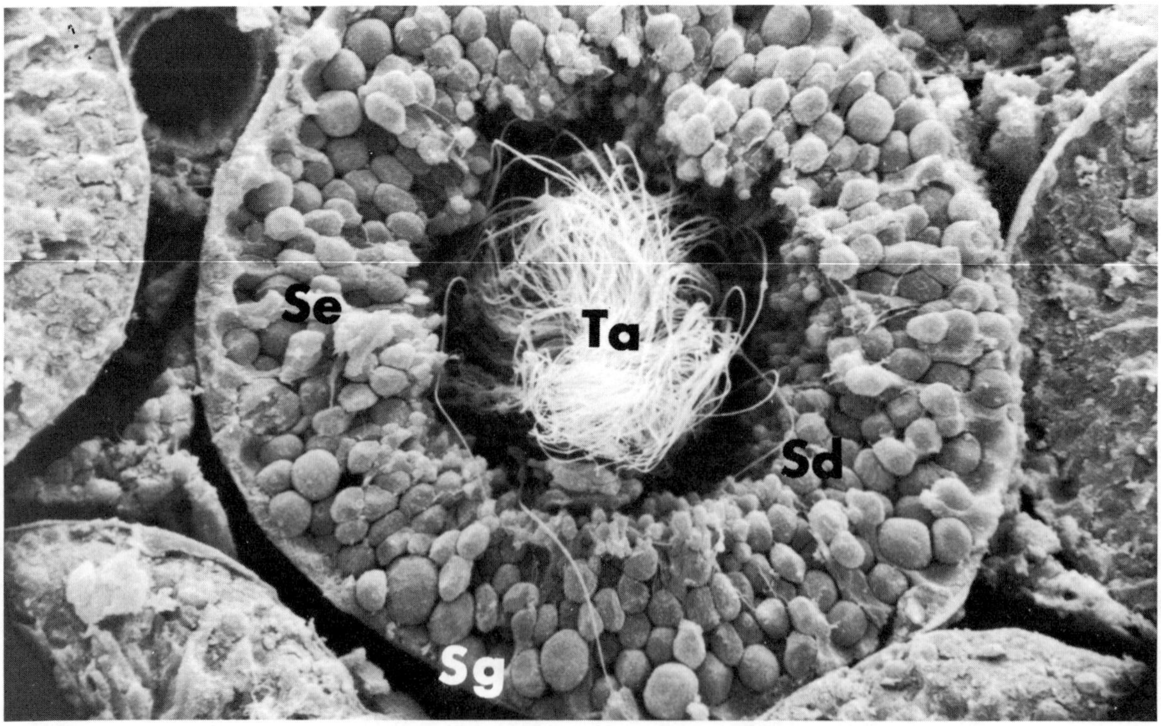

Figure 14-16. Scanning electron micrograph of a seminiferous tubule, 480X. Sc—sercoli cells; Ta—tails of spermatazoa; Sd—spermatids; Sg—spermatagonia.

LABORATORY 15

Structure of the Nervous System

OBJECTIVES

Following this laboratory, you should be able to identify the following structures and state their functional significance.

Gross Anatomy of the Spinal Cord

1. Dura mater
2. Arachnoid
3. Pia mater
4. Central canal
5. Dorsal, ventral, and lateral columns of white matter
6. Dorsal, ventral, and lateral columns of gray matter
7. Dorsal and ventral roots
8. Dorsal root ganglion
9. Dorsal sulci
10. Ventral fissure
11. Dorsal white and gray commissures
12. Ventral white and gray commissures

Gross Anatomy of the Brain

1. Frontal and occipital poles
2. Frontal, parietal, temporal, and occipital lobes
3. Central sulcus
4. Precentral gyrus (primary motor area)
5. Postcentral gyrus (somesthetic area)
6. Primary areas for vision, hearing, and speech
7. Corpus callosum
8. Fornix
9. Thalamus
10. Massa intermedia
11. Aqueduct of Sylvius
12. Cerebellum
13. Fourth ventricle
14. Pons
15. Medulla
16. Pituitary gland
17. Optic chiasma

Blood Supply of the Brain

1. Internal carotid arteries
2. Basilar artery
3. Anterior cerebral arteries
4. Vertebral arteries
5. Circle of Willis
6. Middle cerebral arteries
7. Sagittal sinus

Circulation of the Cerebro-Spinal Fluid

1. Lateral ventricles
2. Third ventricle
3. Fourth ventricle
4. Aqueduct of Sylvius
5. Central canal
6. Arachnoid granulations

Structure of a Neuron

1. Cell body
2. Nucleus
3. Dendrites
4. Axon
5. Nodes of Ranvier
6. Internode
7. Myelin sheath
8. Schwann cells
9. Terminal knobs
10. Synaptic vesicles

Figure 15-1. Diagram of the efferent autonomic pathways.

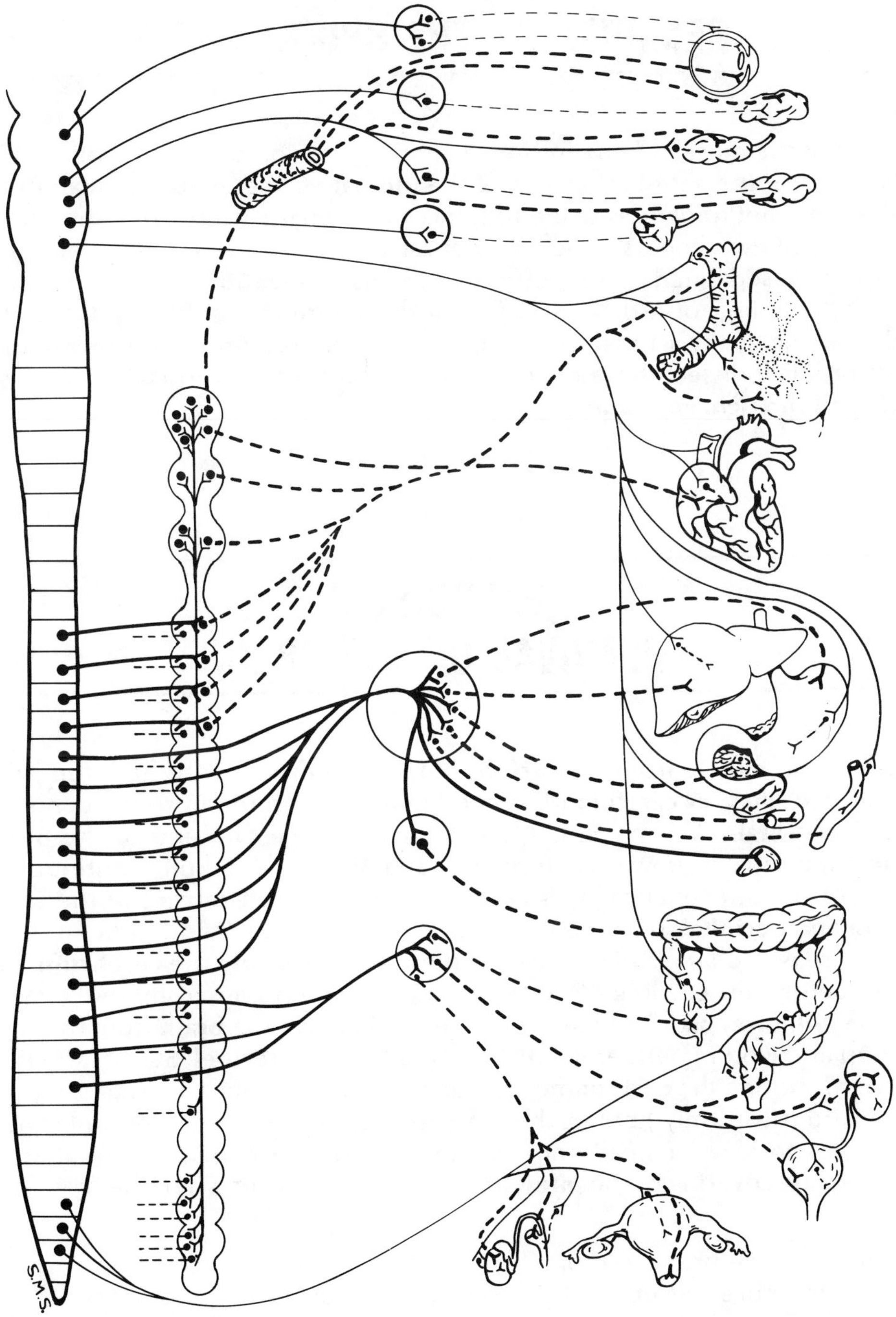

Label one of each of the following:

1. Parasympathetic preganglionic axon.
2. Parasympathetic postganglionic axon.
3. Sympathetic preganglionic axon.
4. Sympathetic postganglionic axon.

EXERCISE 1
REMOVAL OF THE SPINAL CORD

Cut away all of the muscles lying in the vertebral gutter on either side of the back. Using bone clippers, cut the vertebral arches of the lumbar vertebrae just craniad of the root of the tail. Use care not to penetrate the tough connective tissue covering the spinal cord. By placing the tip of the clippers under the neural arch of the vertebrae and prying upward, the bone can be separated from the underlying tissues without injuring them. Continue this process, working toward the head until the entire dorsal half of the spinal cord is exposed. Starting at the caudal end, carefully lift the region of the **sacral** and **brachial plexus**. In order to dissect these networks, it will be necessary to cut through sections of the **pelvis** and **thoracic vertebrae**.

EXERCISE 2
REMOVAL OF THE BRAIN

Clear all skin and muscle from the surface of the cranium. Saw through the **parietal** bone with a small saw. Be very careful not to cut deeply enough to injure the connective tissue covering of the **brain**. When the cut is completed, place one end of the bone clippers under the edge and pry upward, chipping away a piece of the skull. Continue this process until the roof of the cranial cavity is completely removed. Use care in the area over the **cerebellum** and **occipital**, because the bone is especially thick and difficult to remove. In addition, there is a collar of bone between the **cerebellum** and the **cerebrum** that must be taken out before proceeding. Cut the filaments extending from the **olfactory bulbs** through the **cribiform plate** of the **ethmoid bone**. Remove the **brain** from the cranial cavity by lifting up and working under it with the handle of the scalpel. Start at the **medulla** and work forward, cutting the ends of the **cranial nerves** as far from the origins as possible. The **pituitary gland** is so firmly held in the **sulla turcica** that it is removed only with extreme care. It is best to leave a small chip of the floor of the cranial cavity attached to the connective tissue covering of the **pituitary** to be removed later when the brain is free of the cavity.

When removal of the brain and spinal cord has been accomplished, each pair of students should have an entire specimen of the central nervous system, including:

Brain
Cranial nerves
Pituitary gland
Olfactory bulbs

Eye
Spinal cord
Brachial plexus
Sacral plexus

Examples of the:

Spinal nerves
Thoracic region

EXERCISE 3
DISSECTION OF THE SPINAL CORD

Label all structures on Figures 15-2 and 15-3.

1. Notice the lumbar and cervical enlargements of the **spinal cord** and the **conus medullaris** and **filum terminals** at the caudal tip.

2. Observe the **dura mater**, the tough, dense connective tissue covering the **spinal cord** and **brain**. Make a mid-dorsal slit in the lumbar section of the **dura mater** and reflect back the flaps so that the underlying film of the **arachnoid** and **subdural** space can be seen. Locate the fine fibers of the **pia mater** adhering to the surface of the **cord**. Also notice the **denticulate ligaments**, where the **pia** is thickened to attach to the **dura** through the **arachnoid**. These toothlike processes of connective tissue anchor the **dura mater** to sections of the **cord**.

3. Notice the blood supply of the spinal cord. The **anterior spinal artery** occupies the **ventral median fissure**. Where does it arise?

   ____

4. Trace a spinal nerve back toward the **spinal cord.** Cut away the **dura mater,** and observe the **ventral** and **dorsal** roots of the nerve. Locate the **dorsal root ganglion.** If possible in the thoracic region, observe the **white** and **gray rami** connecting the spinal nerve to the **sympathetic ganglion.**

5. Study cross sections of the various levels of the **spinal cord**, and notice the disposition of the **white** and **gray** matter, the **ventral fissure**, the **dorsal median sulcus**, and the **central canal.**

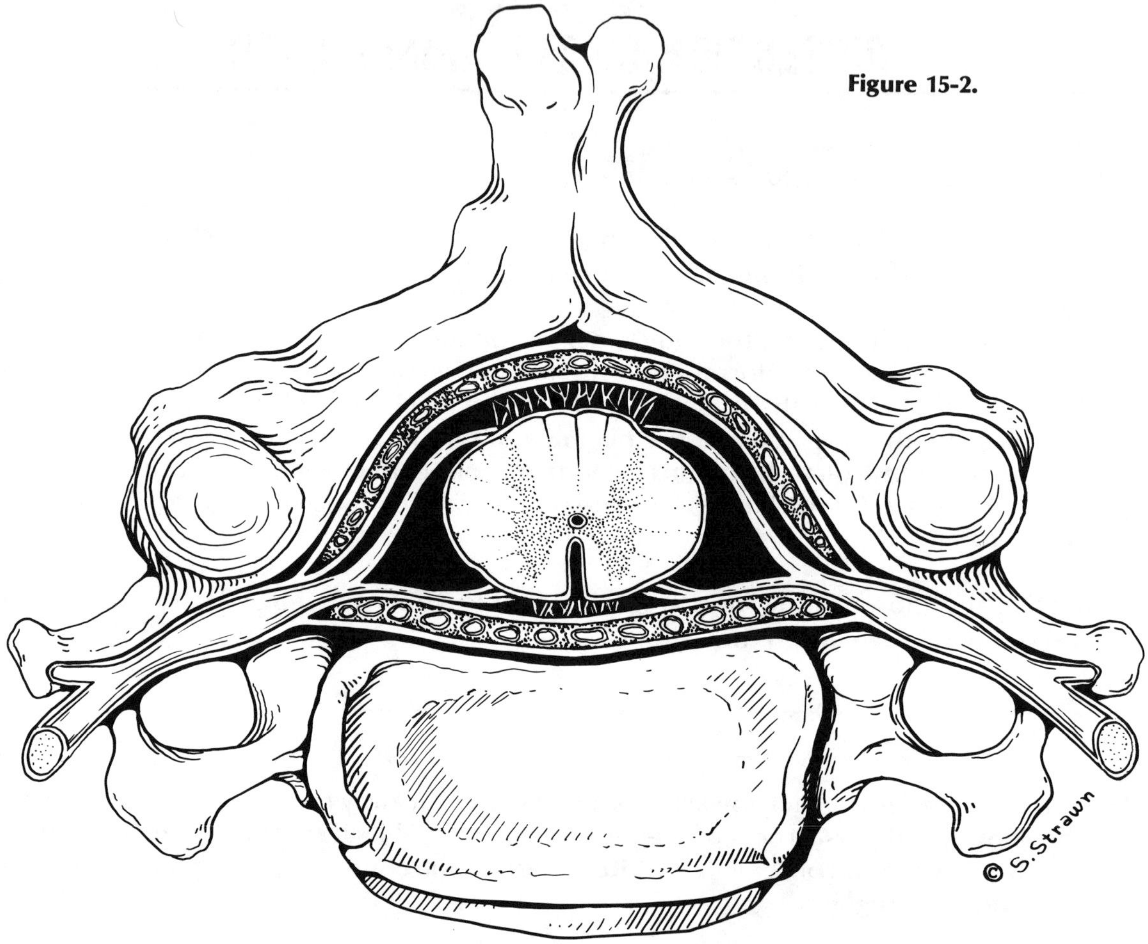

Figure 15-2.

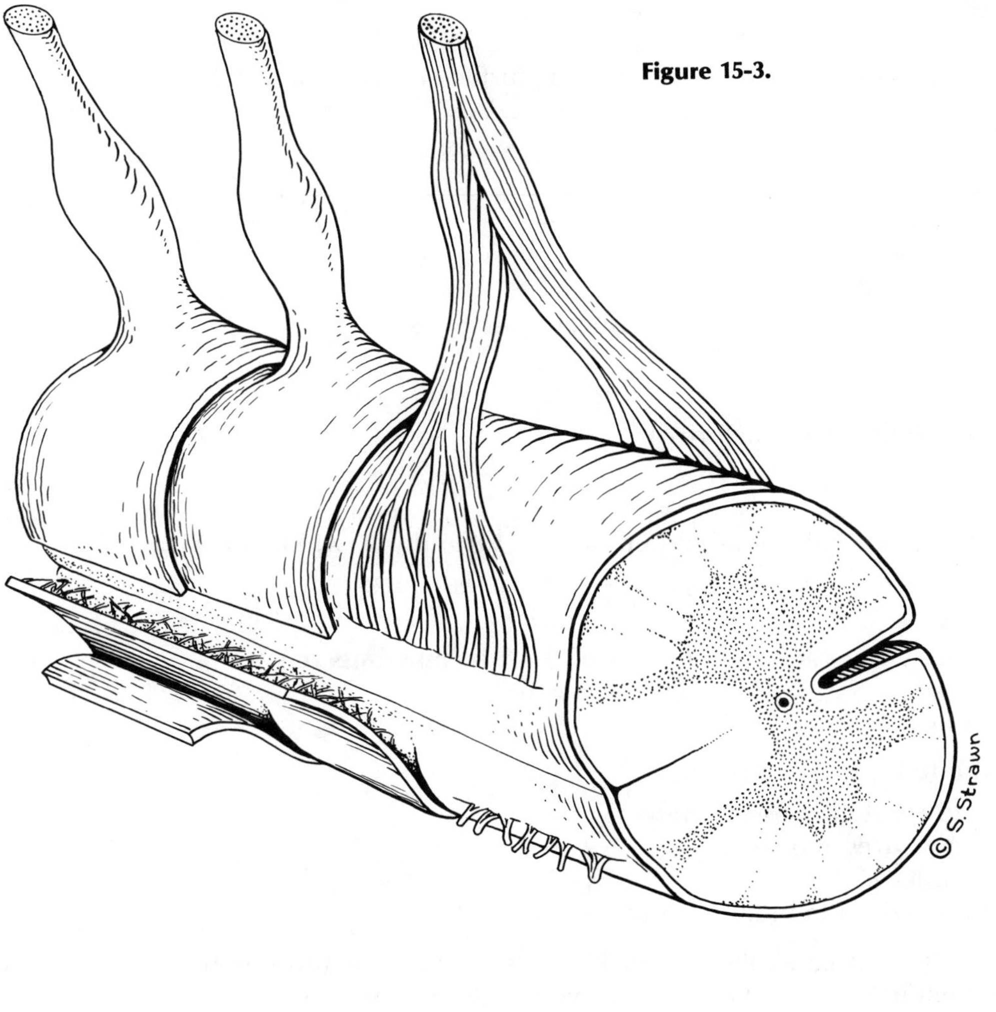

Figure 15-3.

EXERCISE 4
THE BRAIN

Note: Either a cat or sheep brain may be used for this exercise.

1. Observe the cranial **dura mater** and **pia mater**. The **dura mater** extends into the **longitudinal fissure** as the **falx cerebri**.

2. Find the following arteries on the inferior surface of the brain:

   Vertebral
   Basilar
   Anterior spinal
   Posterior cerebral
   Middle cerebral
   Anterior cerebral
   The Circle of Willis and the communicating arteries

3. **Medulla**

   Identify the following structures:

   Pyramids
   Olive
   Cranial nerves arising from the medulla (VI, VII, VIII, IX, X, XI, and XII).

4. **The Pons**

   Identify the sensory and motor roots of the **trigeminal nerve** (V).

5. **Cerebellum**

   Locate the following structures:

   Vermis and **cerebellar hemisphere**
   Transverse fissures and sulci
   The folia or leaves
   Connections with the brain stem (peduncles)

   Remove half of the cerebellum by making a sagittal section through the vermis. Section the removed half horizontally to observe the **dentate nucleus**.

6. **Fourth Ventricle**

   Locate boundaries, communications, and the **choroid plexus**.

   Observe the following structures:

   Corpora quadrigemina
   Cerebral peduncles
   Origin of the **trochlear** (IV) and **occulomotor** nerves (III)

7. The **Inferior Surface**

Locate the following structures that lie close to the **midbrain**:

Optic chiasma
Optic tracts
Pituitary gland
Mammallary bodies

8. **Cerebral Hemispheres**

Observe the fissures, sulci, gyri, and lobes of the hemisphere.

Make a median sagittal section through the brain, and locate the following structures:

Anterior commissure
Corpus callosum
Fornix
Lateral half of the third ventricle
Choroid plexus of the third ventricle
Medial surface of the thalamus
Posterior commissure
Cerebral aqueduct

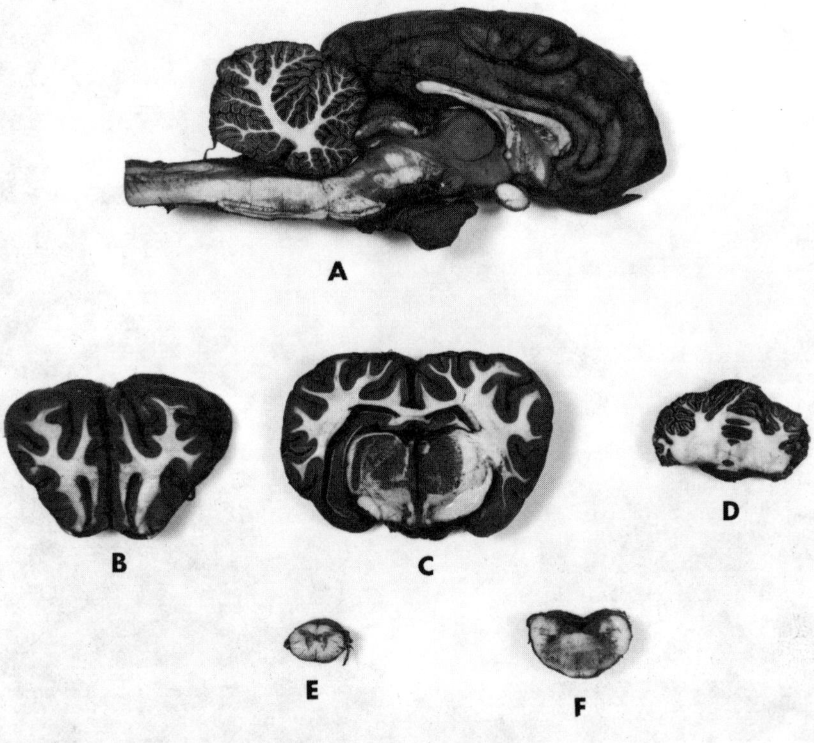

Figure 15-4. Sheep brain stained to show distribution of white and gray matter. A—sagittal section; B—coronal section anterior to corpus callosum; C—coronal section through massa intermedia; D—coronal section through cerebellum; E—cross section of spinal cord; F—section through medulla.

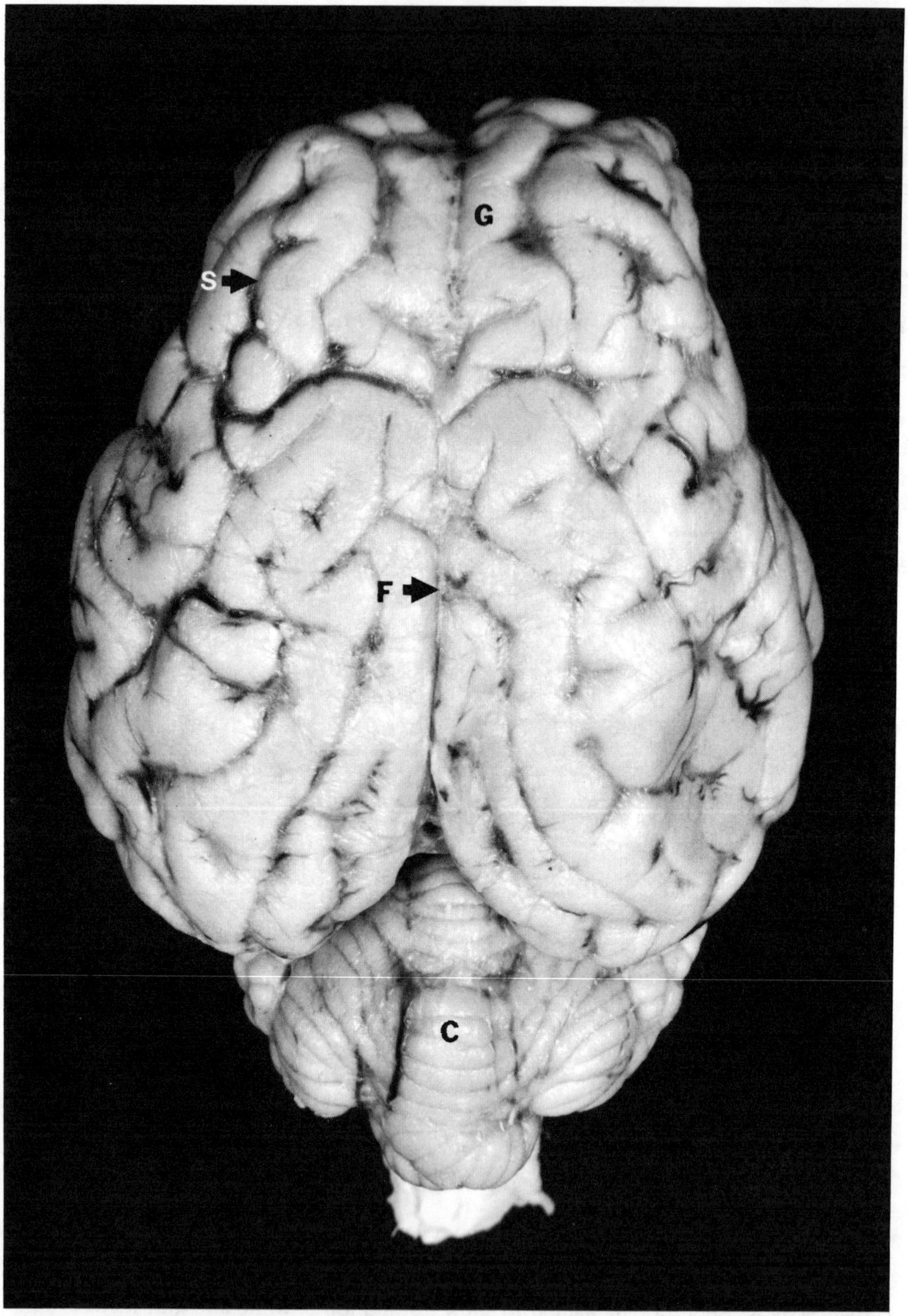

Figure 15-5. Sheep brain, dorsal view. G—gyrus; S—sulcus; F—central longitudinal fissure; C—cerebellum.

Figure 15-6. Sheep brain, dorsal view.

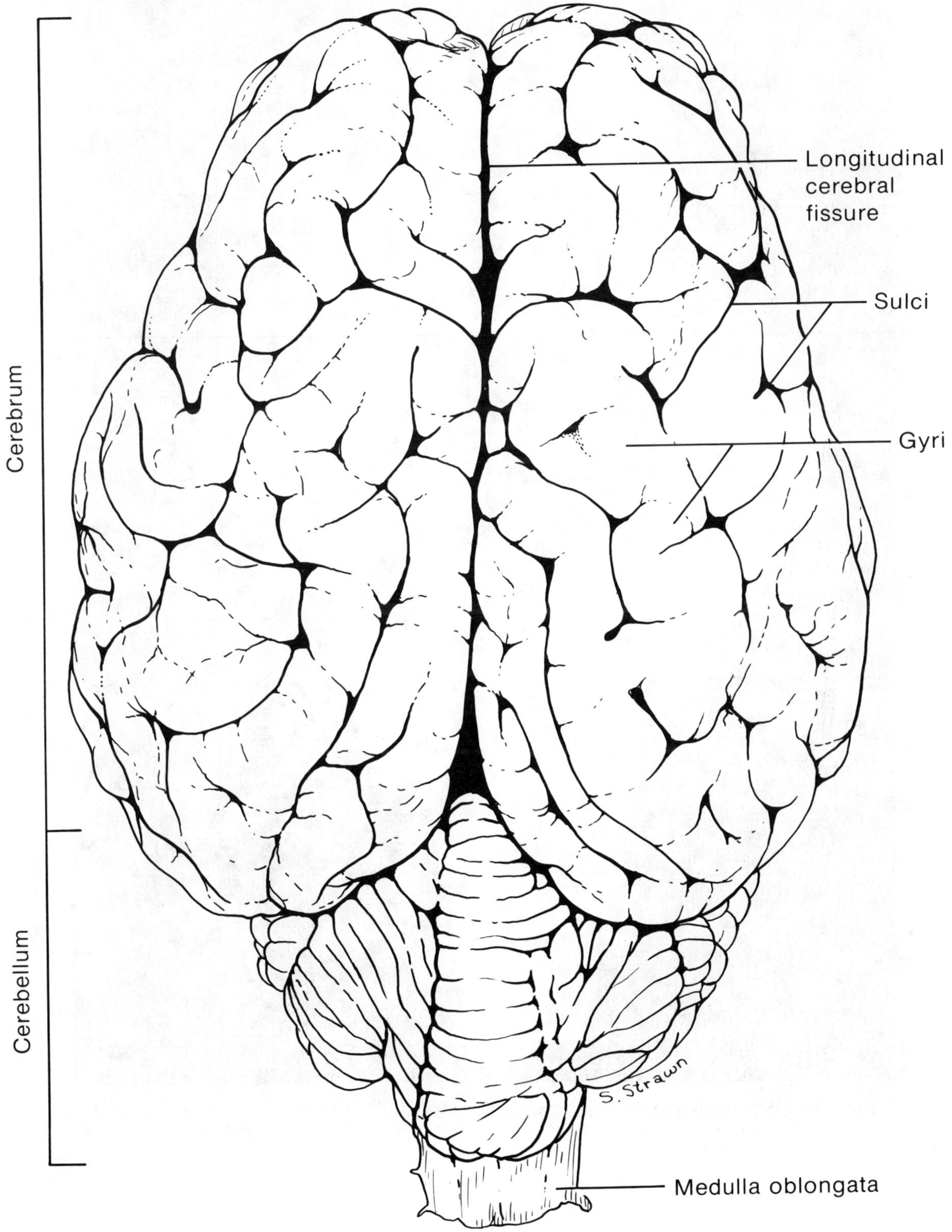

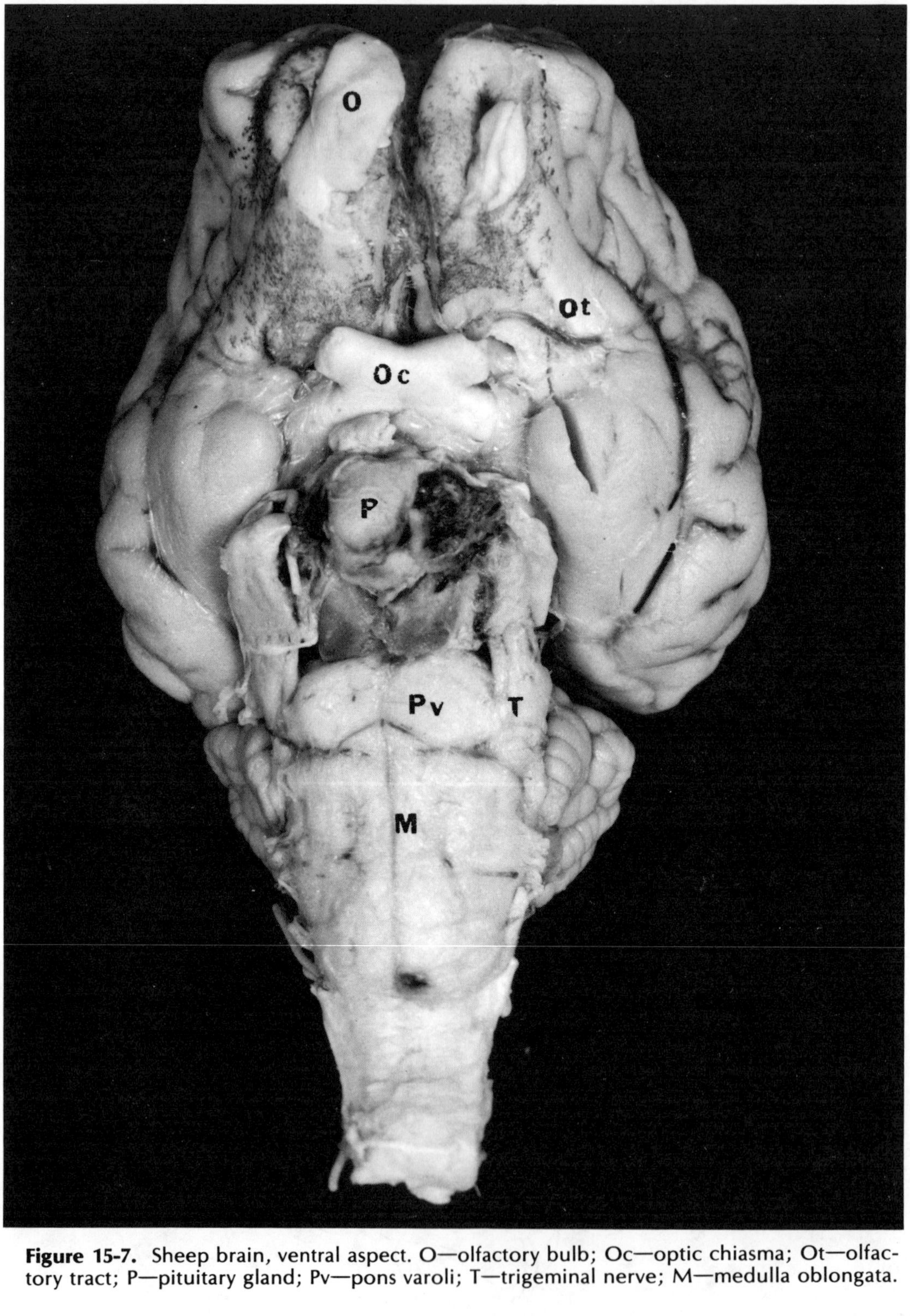

Figure 15-7. Sheep brain, ventral aspect. O—olfactory bulb; Oc—optic chiasma; Ot—olfactory tract; P—pituitary gland; Pv—pons varoli; T—trigeminal nerve; M—medulla oblongata.

Figure 15-8. Sheep brain, ventral view.

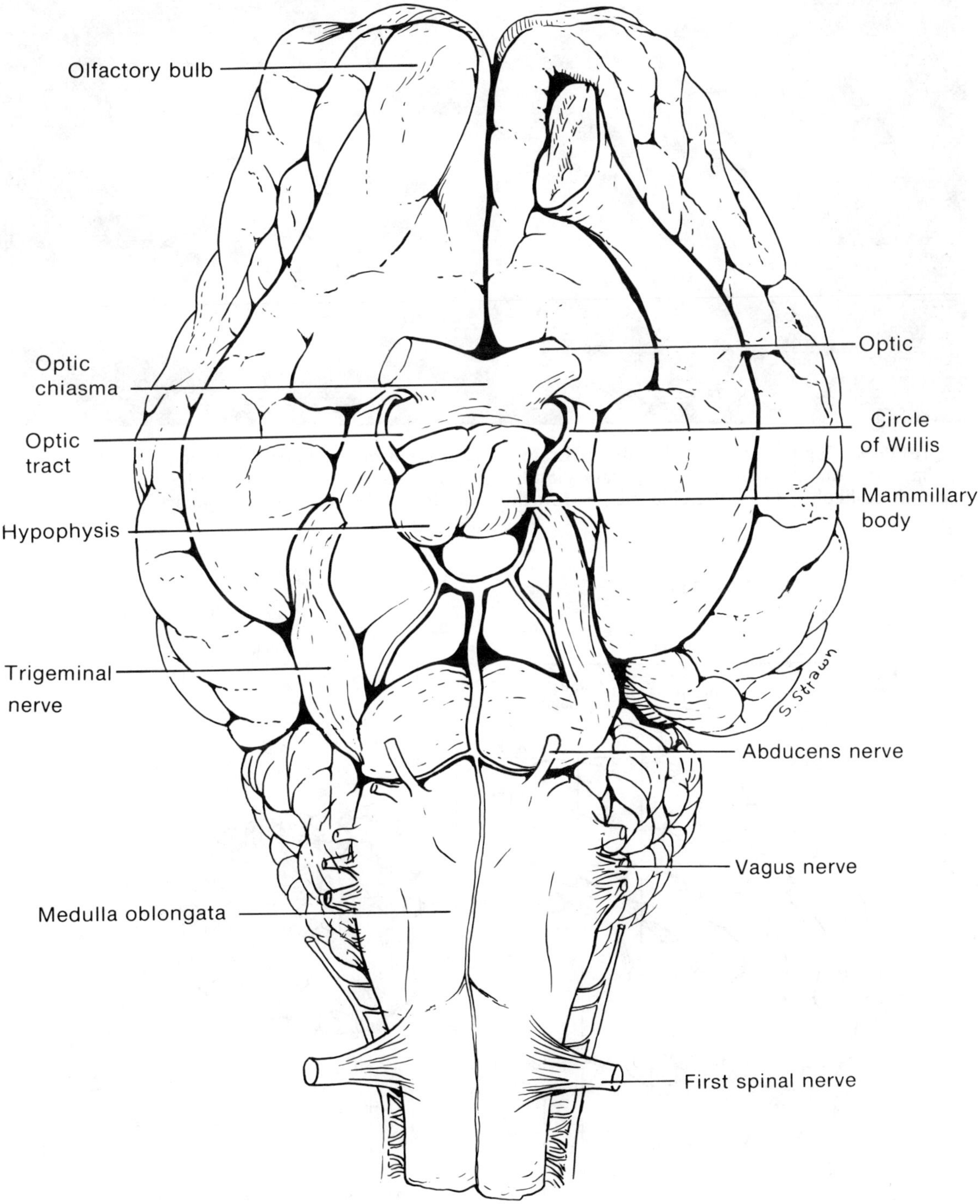

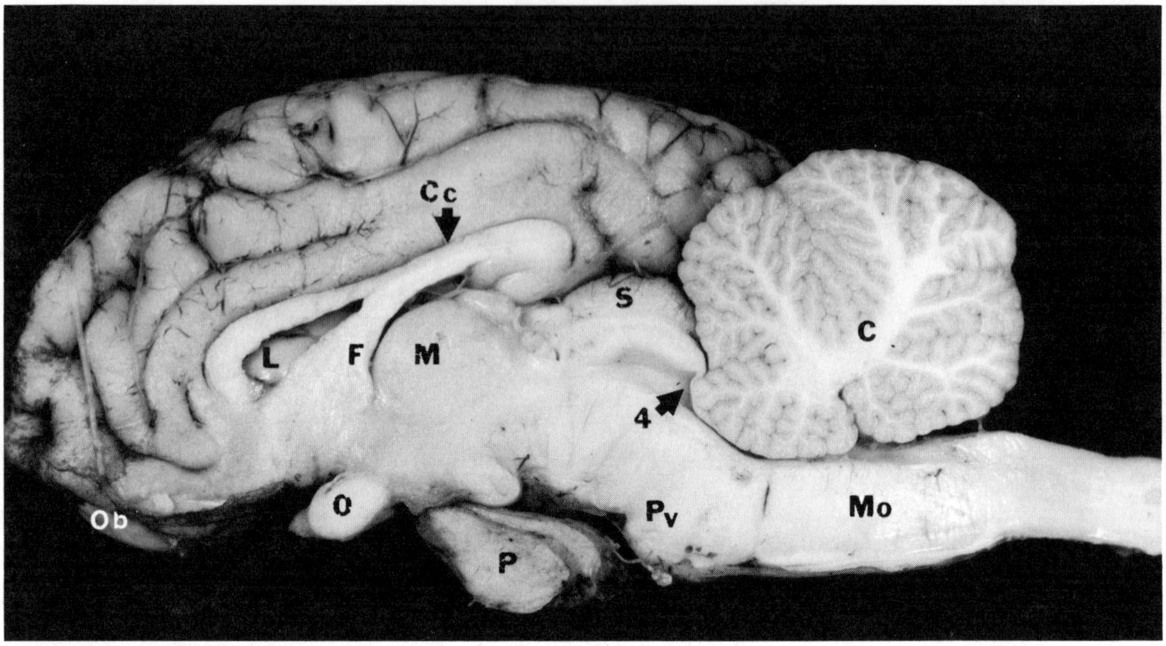

Figure 15-9. Sagittal view of the sheep brain. C—cerebellum; L—lateral ventricle; M—massa intermedia; Ob—olfactory bulb; P—pituitary gland; Pv—pons varoli; Cc—corpus callosum; F—fornix; S—superior colliculus; O—optic chiasma; 4—fourth ventricle; Mo—medulla oblongata.

Figure 15-10. Sheep brain, sagittal view.

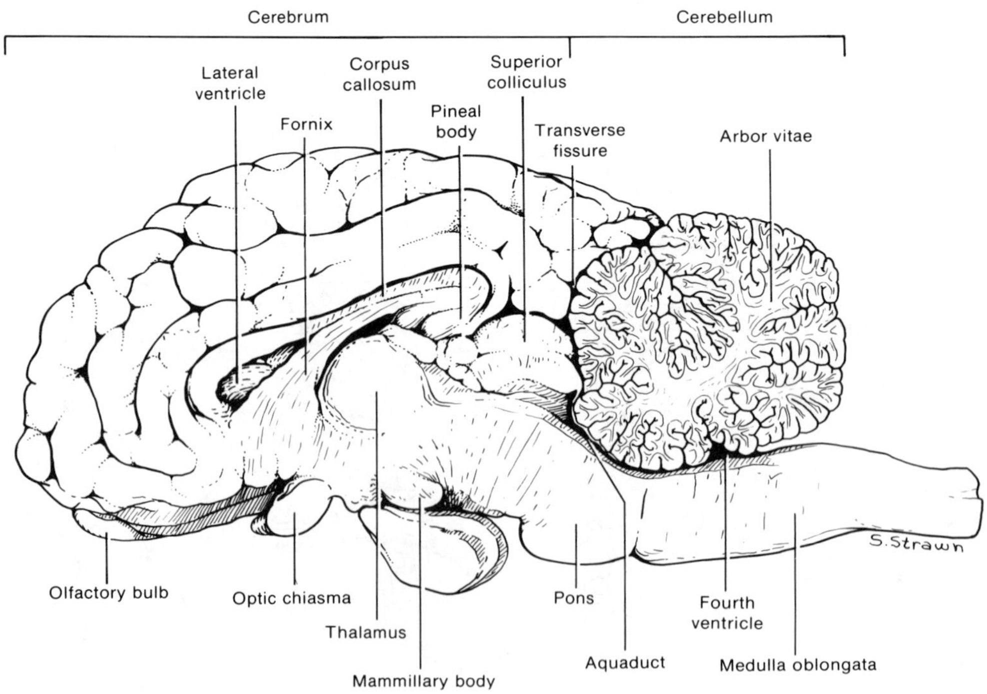

Table 15-1

Cranial Nerves of the Cat

Nerve		Function	Cranial Exit	Origin	Termination
I	Olfactory	Special sensory	Cribiform plate	Olfactory bulb	Olfactory epithelium
II	Optic	Special sensory	Optic foramen	Optic chiasma	Retina
III	Occulomotor	Motor	Sphenoidal fissure	Crura cerebri	Inferior oblique medial, superior, inferior rectus muscles of eye
IV	Trochlear	Motor	Sphenoidal fissure	Mesencephalon muscle	Superior oblique eye
V	Trigeminal	Mixed		Medulla	
	a. Ophthalmic	Sensory	Sphenoidal fissure		Nose, upper eyelid, skin of cranium, tear glands
	b. Maxillary	Sensory	Foramen		Upper teeth, lips, cheek
	c. Mandibular	Mixed	Foramen ovale		Motor to muscle of chewing, skin. Sensory to lower teeth
VI	Abducens	Motor fissure	Sphenoidal	Trapezoid body	Lateral rectus eye muscle
VII	Facial	Motor	Internal auditory meatus	Trapezium	Muscle of facial expression, digastric, sub-max, and subling glands
VIII	Auditory	Special sensory	Internal auditory meatus	Auditory eminence	Labyrinth
IX	Glosso-pharyngeal	Sensory, motor, special sense (?)	Jugular foramen	From medulla with X	Pharynx and soft palate, taste buds, parotid gland
X	Vagus	Motor and sensory	Jugular foramen	Later portion of medulla	Pharynx, trachea, larynx, esophagus, heart, lungs, and upper abdominal viscera
XI	Spinal accessory	Chiefly motor	Jugular foramen	Medulla	Trapezius and the sternomastoideus
XII	Hypoglossal	Chiefly motor	Hypoglossal foramen	Posterior level of the IV ventricle	Muscles of the tongue

EXERCISE 5
COMPARISON OF CAT OR SHEEP AND HUMAN BRAINS

On the model of the human brain, and on the dissected human brain on the demonstration table, locate as many of the above named structures as possible. Compare structures of the human brain with those of the cat or sheep.

In the space below and on the next page, discuss the differences between the brain of the cat or sheep and the brain of the human, and relate them to behavioral differences.

LABORATORY 16

Anatomy and Physiology of the Eye and Ear

OBJECTIVES

By the end of this laboratory period, you should be able to identify the following structures on dissections or models.

Gross Anatomy of the Eye

Identify the following structures on drawings, diagrams, or dissections, and state their functional significance.

1. Sclera
2. Choroid
3. Retina
4. Cornea
5. Iris
6. Ciliary body
7. Suspensory ligaments
8. Lens
9. Fovea centralis
10. Optic nerve
11. Superior rectus
12. Medial rectus
13. Lateral rectus
14. Inferior rectus
15. Superior oblique
16. Inferior oblique

Aiming the Eye

Describe the actions of the extrinsic muscles of the eye in moving the eyeball, and be able to state their intervention by the following nerves:

1. Occulomotor (III)	Superior, medial, and inferior rectus, and inferior oblique
2. Trochlear (IV)	Superior oblique
3. Abducens (VI)	Lateral rectus

Anatomy of the Ear

1. Pinna
2. External auditory meatus
3. Tympanic membrane
4. Malleus
5. Incus
6. Stapes
7. Eustachian tube
8. Cochlea
9. Semicircular canal
10. Vestibule

EXERCISE 1
ANATOMY OF THE CAT EYE

Procedure

1. In your previous dissection of the brain, you were instructed to remove, along with the brain, at least one eye with its nerves and muscles attached. Your instructor may prefer to substitute a sheep or cow eye.

2. Clear the fat and fascia from the surface of the eyeball so that the **extrinsic muscles** and the **optic nerve** are exposed. Locate the following structures:

   a. Lateral rectus
   Medial rectus
   Superior rectus
   Inferior rectus

   These muscles are named from the fiber direction and from their position on the surface of the eyeball. All originate from the **sphenoid bone** around the margin of the **orbital cavity**. They insert on the **sclera**. All are innervated by the **occulomotor nerve** (III), except the **lateral rectus**, which is served by the **abducens** (VI). Find the nerves as they enter the muscles.

   b. The **superior oblique** originates on the cranial border of the orbit and runs up and back to a small pulley, where it changes direction and runs down and laterally to insert on the **sclera**, just caudal to the insertion of the **superior rectus**. Identify this muscle, and locate the **trochlear nerve** (IV) that innervates it.

   c. The **inferior oblique** originates from the orbital surface of the **maxilla**, lateral to the **lacrimal bone**. It curves over the ventral surface of the eyeball, crosses the tendon of insertion of the **inferior rectus**, and inserts into the **sclera** under the ventral margin of the insertion of the **lateral rectus**. It is innervated by the **occulomotor** (III).

   d. The **ciliary ganglion** is a triangular mass of nervous tissue that lies on the dorsal surface of the **inferior rectus** muscle. Cut and reflect back the **retractor oculi**, which lies between the **optic nerve** and the **inferior rectus**. This will expose the entire **ciliary ganglion**. It is then possible to trace the **short postganglionic**.

Figure 16-1. Structure of the eye, retina, fovea, and ciliary body.

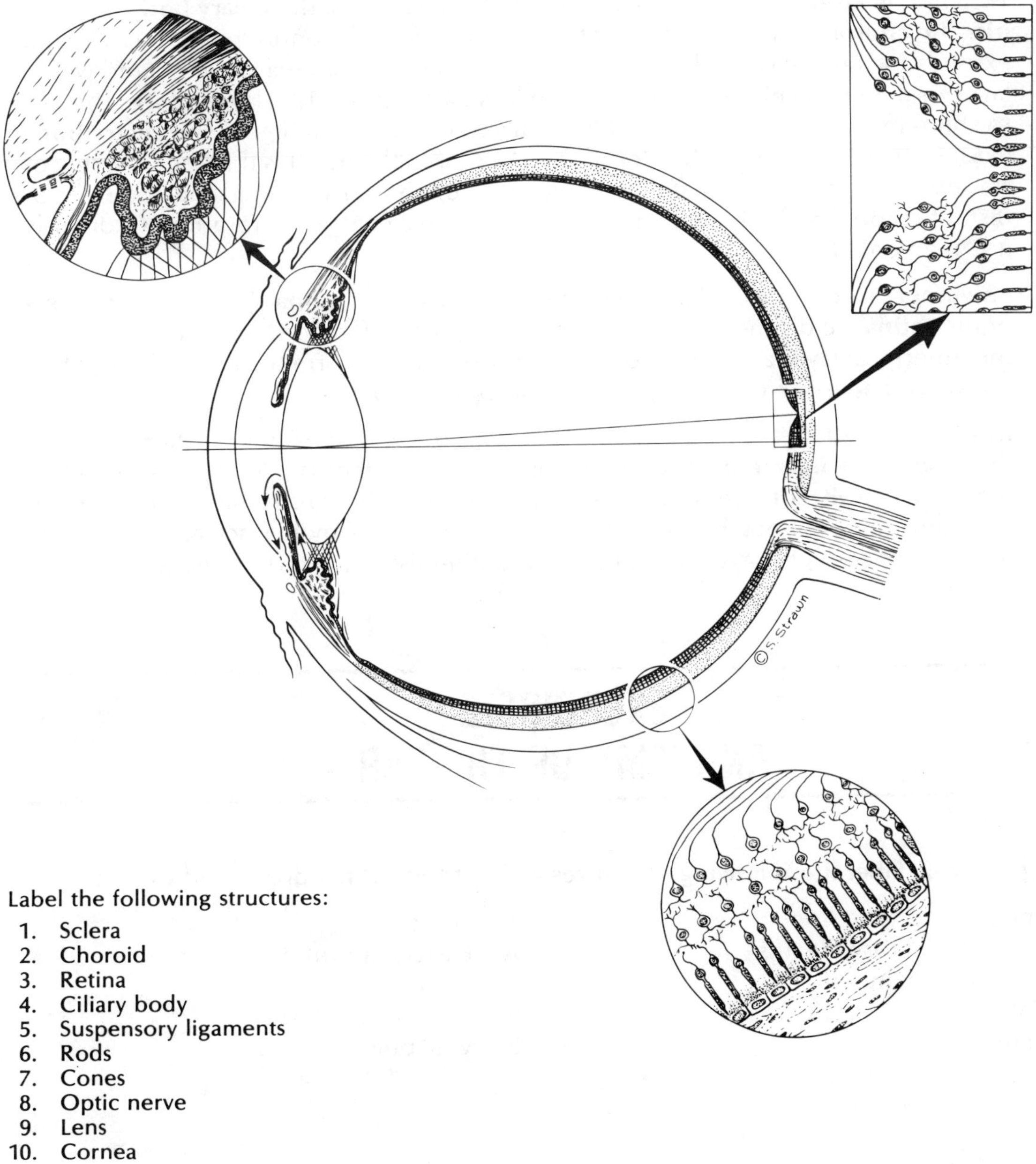

Label the following structures:

1. Sclera
2. Choroid
3. Retina
4. Ciliary body
5. Suspensory ligaments
6. Rods
7. Cones
8. Optic nerve
9. Lens
10. Cornea
11. Iris

3. When all extrinsic muscles and nerves have been located, clear them from the eyeball, and dissect the eyeball according to the following instructions.

   a. Bisect the eyeball, and place it under water to observe the structures listed below. Label the following structures on Figure 16-1.

      The **sclera** and **cornea** are the two sections of the tough, fibrous protective outer coat of the eyeball. The **sclera** is white and opaque and surrounds the larger posterior portion of the eyeball, while the **cornea** is transparent and has a smaller radius of curvature, which causes it to form a bulge in the anterior surface.

      The middle layer or vascular coat consists of the **choroid**, the **ciliary body**, and the **iris**. The **choroid** is the largest part of this layer, covers the entire posterior half, and extends forward beyond the "equator." In carnivorous animals, it is coated with a layer of flattened cells referred to as the **lamina tapetum**. This layer causes the eye to shine in the dark and is probably an aid to vision in dim light. The **ciliary body** extends from the anterior border of the **choroid** to the **iris**. It is thickened, a ring of smooth muscle that regulates the thickness of the lens in focusing for near or distant vision. The **iris** is a thin disc of muscle with a central perforation. Ordinarily it is pigmented.

      The inner layer of the eyeball is the nervous coating, or **retina**. Although ten layers of the **retina** are distinguishable in a microscopic preparation, only the fibers form the innermost layer and can be observed grossly. These form the layer of fibers that converge and exit from the eyeball as the **optic nerve**.

      Inside the eyeball, the **lens** can be observed, suspended from the **ciliary body** by the **suspensory ligaments**. The cavity posterior to the lens is filled with a refracting medium, the **vitreous humor**. The cavity anterior to the lens is filled with a similar medium, the **aqueous humor**. These three refracting media act as a system of lenses to focus light rays on the retina in a diminished, inverted image.

EXERCISE 2
ANATOMY OF THE EAR

On Figure 16-2, label the following structures and list their functional significance:

1. Pinna
2. External auditory meatus
3. Tympanic membrane
4. Malleus
5. Incus
6. Stapes
7. Eustachian tube
8. Cochlea
9. Semicircular canal
10. Vestibule

Figure 16-2. The human ear.

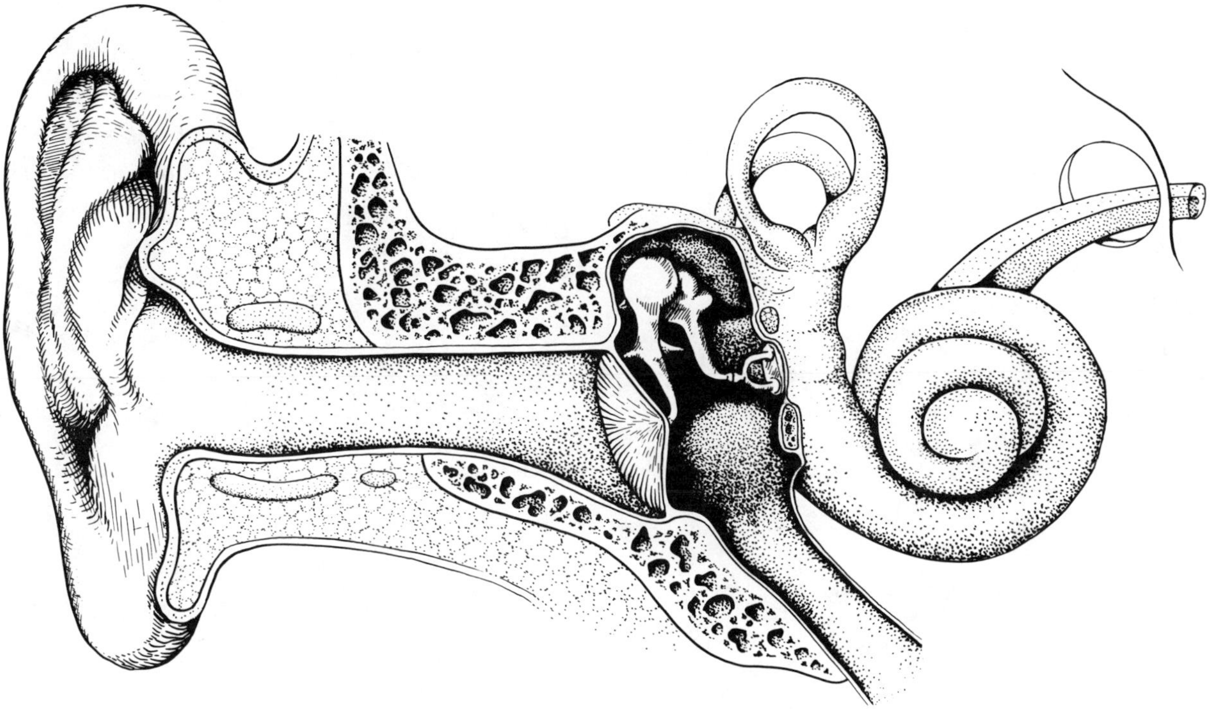

1. Pinna ______
2. External auditory meatus ______
3. Tympanic membrane ______
4. Malleus (incus and stapes) ______
5. Eustachian tube ______
6. Cochlea ______
7. Semicircular canals ______
8. Vestibule ______